Ship and Mobile Offshore Unit Automation

Ship and Mobile Offshore Unit Automation

A Practical Guide

Henryk Pepliński

Gulf Professional Publishing
An imprint of Elsevier

Gulf Professional Publishing is an imprint of Elsevier
50 Hampshire Street, 5th Floor, Cambridge, MA 02139, United States
The Boulevard, Langford Lane, Kidlington, Oxford, OX5 1GB, United Kingdom

Notices
Knowledge and best practice in this field are constantly changing. As new research and experience broaden our understanding, changes in research methods, professional practices, or medical treatment may become necessary.

Practitioners and researchers must always rely on their own experience and knowledge in evaluating and using any information, methods, compounds, or experiments described herein. In using such information or methods they should be mindful of their own safety and the safety of others, including parties for whom they have a professional responsibility.

To the fullest extent of the law, neither the Publisher nor the authors, contributors, or editors, assume any liability for any injury and/or damage to persons or property as a matter of products liability, negligence or otherwise, or from any use or operation of any methods, products, instructions, or ideas contained in the material herein.

Library of Congress Cataloging-in-Publication Data
A catalog record for this book is available from the Library of Congress

British Library Cataloguing-in-Publication Data
A catalogue record for this book is available from the British Library

ISBN: 978-0-12-818723-4

For information on all Gulf Professional publications
visit our website at https://www.elsevier.com/books-and-journals

Publisher: Brian Romer
Senior Acquisition Editor: Katie Hammon
Editorial Project Manager: Joanna Collett
Production Project Manager: Sruthi Satheesh
Cover Designer: Mark Rogers

Typeset by SPi Global, India

Contents

Preface

This book is based on personal experience gained over 40 years in the marine, oil and offshore industries. During these years I was designing electrical and automation systems while working in a shipyard, approving electrical and automation systems while working for classification societies, commissioning electrical and automation systems in different shipyards and supervising ship and offshore unit conversions while working in Owner's technical offices.

Notwithstanding, I apologize to those who may find the book does not include all information on ship and mobile offshore unit automation. The subject is very wide and dynamic and I would therefore appreciate any constructive comments which I may take into consideration when considering a future revision of the book.

Henryk Pepliński

Acknowledgements

The author wishes to thank and gratefully acknowledge all those who provided material and advice for the production of this book, particularly the following:

James MacGregor, former Chief Technical Officer, Prosafe Offshore, for advice on scope of the book and book review.

Radosław Sochanowski, Kongsberg Maritime Poland Managing Director, Kalina Kielbratowska, Marek Gośniak, Mariusz Dąbrowski and Jacek Jankowski for provision of Kongsberg documentation and technical assistance.

Adam Woźniewski, Wartsila Ship Design Poland Sp. z o.o. for provision Wartsila documentation and technical assistance.

Petter Bertzen ABB AS Marine & Ports and Piotr Białek ABB Sp. z o.o. for technical assistance.

Chapter 1

Preamble

In the first half of the twentieth century seagoing vessel technical systems and processes were relatively simple and all were controlled manually and monitored by the Marine Engineer Watchkeeper. From the beginning of the 1960s, onboard equipment became more complicated and new vessels were equipped with electric, electronic, pneumatic and hydraulic automation equipment to control and monitor dedicated system processes remotely. These control systems continued to develop into the sophisticated software based systems nowadays installed on most vessels and offshore units.

Nowadays, due to limited manpower on vessels, high crew costs and rapid development of computers and computer based control and monitoring systems, most onboard processes are remotely controlled and monitored from the Bridge as well as from the Engine Control Room (ECR). A significant percentage of vessel equipment is automated and can work without crew assistance in automatic mode for a long time.

This book is written for marine Electrical Engineers, Electrical and Instrumentation Superintendents, ship designers and university students, and outlines in detail the current status of control and monitoring systems on board both conventional vessels and offshore units.

The final chapter of the book describes future trends in development of control and monitoring systems, including usage of Artificial Intelligence (AI) in software development, and prospects for use of autonomous vessels.

Ship and Mobile Offshore Unit Automation. https://doi.org/10.1016/B978-0-12-818723-4.00001-6

Chapter 2

General

Abstract

Industrial control and monitoring systems are developing rapidly and have a big influence on the maritime and oil and gas industries. This development process for automation systems requires not only permanent update of knowledge by designers of ship and offshore unit automation systems, but also seagoing Marine Engineers, Electrical and Instrumentation Engineers (EEN), Electronic Technical Officers (ETO) and onshore based Electrical and Instrumentation Superintendents. It is also necessary that maritime engineering university students receive knowledge in modern automation systems. Moreover, specialists involved in all facets of vessels and offshore unit control and monitoring systems design, installation, commissioning, maintenance, operation, and surveying should continuously update their knowledge from published literature, international standards, equipment/systems manuals and by attending courses provided by control and monitoring systems manufacturers, and technical and maritime educational institutes. It is recommended that personnel involved with these systems are notified by responsible staff at the companies and educational institutions of any new or modified maritime legislation and updated guidance related to ship and offshore unit control and monitoring systems development.

This book contains some background information necessary for specialists involved in development, installation and operation of control and monitoring systems for ships and offshore units. The book provides information and guidance for vessel and offshore unit personnel who deal with control and monitoring systems onboard or ashore.

In the beginning of the book information is provided defining the concepts of manned and unmanned Engine Control Room (ECR) and Bridge. These definitions are based on international statutory requirements from the International Maritime Organization (IMO) such as Safety of Life at Sea Convention (SOLAS Convention), Code for the Construction and Equipment of Mobile Offshore Drilling Units (MODU Code) and rules from classification societies such as Lloyds Register (LR), American Bureau of Shipping (ABS), and Det Norske Veritas Germanischer Lloyd (DNVGL).

The following chapters of the book describe the requirements for control and monitoring systems contained in International Maritime Organization (IMO) publications: SOLAS Convention, MODU Code and Classification Societies rules from American Bureau of Shipping (ABS) and Det Norske

Veritas Germanischer Lloyd (DNVGL). Other Classification Societies Rule requirements, for example: Rules of Lloyds Register (LR), Bureau Veritas (BV), China Classification Society (CCS), Nippon Kaiji Kuokai (NK), Indian Register of Shipping (IRS), Russian Maritime Register of Shipping (MRS), Polish Register of Shipping (PRS), Registro Italiano Navale (RINA), Croatian Register of Shipping (CRS) may be found in their publications available on their web pages.

The next chapters of the book outline other guidance and requirements applicable to marine and offshore control systems. These requirements and guidance are found in International Electro-technical Commission (IEC) Standards, Institute of Electrical and Electronics Engineers (IEEE) Standards, European Union (EU) Directives, American Petroleum Institute[1] (API) Recommended Practices and Standards, and Norwegian Maritime Authority (NMA) rules and NORSOK Standards[2]. The above listed standards are followed by some vessel and offshore unit owners. The additional requirements in these standards are often based on conclusions from earlier marine accidents. Some subject addressed include the development of an Alarm Management Philosophy and concept of Safety Integrity Level (SIL) for safety systems on Floating, Production, Storage and Offloading (FPSO) vessels and other offshore units.

A subsequent chapter is intended to assist the reader to distinguish between obligatory regulations such as those included in maritime Conventions and Codes signed by Flag States (for example SOLAS, MODU Code etc.) and application of different Classification Societies Rules required to obtain a vessel or offshore unit Class certificate. Later in that chapter an explanation is provided as to the necessity, or otherwise, of implementing European Union Directives, and international standards such as International Electro-technical Committee (IEC), International Organization for Standardization (ISO) and additional specific maritime administration requirements.

These chapters introducing the applicable rules and regulation are followed by a chapter describing the activities involved in designing control and monitoring systems for ships and offshore units, introducing the commonly used SFI marine engineering drawing numbering system.

The steps involved in specifying and purchasing control and monitoring systems are presented. Descriptions are provided of the various stages including Vendor Quotation, Information to Tender (ITT) and Request for Quotation (RFQ). In this chapter are included examples of Technical Agreement (TA), Systems Functional Descriptions (FD) and a few P&ID and MIMIC diagrams.

---

1. API webpage http://www.api.org/.
2. https://www.standard.no/en/sectors/energi-og-klima/petroleum/norsok-standards/#. W89PB2dDtis.

The next chapters contain descriptions of typical vessel and offshore unit Integrated Automation Systems with associated subsystems such as: Auxiliary Control System, Alarm and Monitoring System and Power Management System and their interconnections. Additionally, examples are provided of typical Alarm and Monitoring lists, print screens of Alarms on Bridge Operator Station and Main Power Distribution System Single Line Diagrams (SLD).

These chapters are followed by descriptions of Safety Management Systems including Fire & Gas Detection systems, with examples of Fire and Gas Philosophy, Cause & Effect diagrams and Emergency Shutdown system supplemented by Emergency Shutdown Philosophy. Additionally in this chapter is presented the Public Address/General Alarm System including a typical block diagram.

Thereafter position keeping systems are described, including Dynamic Positioning Systems and Position Mooring Systems using anchors. In this chapter examples are provided of machinery automation systems integrated with Dynamic Positioning systems.

Other chapters outline certification principles for control and monitoring systems and requirements and guidance for practical installation of control and monitoring systems. Typical documents used to perform installation of control and monitoring systems in controlled conditions are described; e.g. shipyard standard practices, system producers commissioning instructions. Additionally, installation guidelines for automation system elements and Electromagnetic Compatibility considerations of equipment installation are provided.

Requirements and procedures for commissioning control and monitoring systems are described, including the mechanical completion phase and including associated documentation, certificates, check records, punch lists and handover process.

Different methods to supervise maintenance of control and monitoring system elements are introduced, including Planned Maintenance Systems (PMS) and Computerized Maintenance Management Systems (CMMS).

Requirements for vessel/offshore unit certification are introduced including Health and Safety Executive requirements and guidance applicable offshore on the United Kingdom continental shelf (UKCS), and Norwegian offshore requirements and guidance published by the Norwegian Maritime Authority (NMA) and the Petroleum Safety Authority (PSA), as well as International Marine Contractors Organization (IMCA) guidelines.

The evolution of control and monitoring systems from the 20th century until the present day is outlined. This chapter additionally outlines recent and projected developments in automation of autonomous ships and the needs of development for such ships, including advanced redundant sensors. Also described are autonomous ships control concepts Shore Control Centre Operator, Shore Control Centre Engineer and Shore Control Centre Situation Team. New trends to use Artificial Intelligence (AI) onboard in applications such as Decision Support Systems (DSS) and building AI-models to fully Integrated Automation Systems (IAS) are discussed, together with Cyber security threats.

The Appendices contain more detailed information which may be useful for readers such as;

Appendix 1A—List of contract regulations, requirements and standards usually included in ship technical specifications. This appendix lists contract regulations, requirements and standards often quoted in the contract between shipyard and ship owner including Classification Rules, National Rules and Regulations, International Rules and Regulations i.e. IMO Conventions, IMO Codes, other technical requirements, IMO Resolutions and Special Rules and regulations.

Appendix 1B—List of contract regulations, requirements and standards usually included in offshore unit technical specification. This appendix lists contract regulations, requirements and standards often quoted in the contract between shipyard and offshore unit owners including Classification Rules, Regulatory Requirements and Industrial Codes and Standards.

Appendix 2—Automation Design Examples. This appendix includes different examples of different phases of the vessel or offshore unit designing process. For the basic design process are presented extracts of Control and Monitoring Elements, Arrangement in Engine Room, automation systems descriptions i.e. Automatic Vessel Control Description, Power Management System (PMS) Description, Alarm and Monitoring System Description. Included for the detailed engineering design phase are documents such as Wiring Diagrams, examples of IAS Termination list, I/O list and IAS Control and Monitoring Cables Numbering Principles. Technological documentation examples include Installation Instructions for control and monitoring cables, Work instructions for Structured Network – Test and Termination.

FIG. 2.1 *Safe Notos* DP floatel working in Brazilian waters with IAS, SAS, PMS and DP systems. *Picture received from Prosafe AS, Norway.*

Appendix 3—Web References—This appendix lists web addresses referenced in the book. Web references are presented for each chapter separately.

Appendix 4—International Electro-technical Committee (IEC) Standards relevant to ship and mobile offshore automation systems. This appendix lists many IEC Standards that are often referenced in Technical specification and documentation from control and monitoring systems suppliers.

Appendix 5—Abbreviations

In Fig. 2.1 is presented Prosafe Accommodation vessel *Safe Notos* with IAS, Safety Automation Systems and Position keeping systems working in Brazilian waters.

Chapter 3

Type of operation, engine room and bridge

Chapter outline

Abstract

The required mode of engine room and bridge operation has a big influence on control and monitoring system scope and consequently the actual operation of the vessel. In terms of monitoring machinery and equipment engine room can be manned or unmanned during a voyage. This chapter explains the differences between manned and unmanned engine room and bridge operation based on definitions from International Maritime Organization (IMO) Safety of Life at Sea Convention (SOLAS Convention), IMO Code for the Construction and Equipment of Mobile Offshore Drilling Units (MODU Code) and three classification societies, i.e., Lloyds register (LR), American Bureau of Shipping (ABS), Det Norske Veritas Germanischer Lloyd (DNVGL). Other classification societies rules requirements for manned and unmanned engine room and bridge can be found in their published rules.

3.1 Manned engine room

Ships with permanently manned engine rooms are rare nowadays. Because the environment in an engine room is usually hot, humid and noisy, a normally unmanned engine room is a considerable benefit for the engineering crew. However, it could still be feasible to have manned engine rooms if engine room processes and electrical installations are kept simple. The IMO SOLAS Convention[1] includes basic requirements for manned engine rooms, for example: automatic starting of stand by generator sets, automatic closing of dead bus bar after failure of the running generator and automatic starting of essential electric consumers.

---

1. SOLAS Convention—IMO Safety of Life at Sea Convention.

Ship and Mobile Offshore Unit Automation. https://doi.org/10.1016/B978-0-12-818723-4.00003-X

3.2 Unmanned engine rooms

The IMO SOLAS Convention incorporates additional requirements for ships with periodically unattended machinery spaces including: fire precautions, protection against flooding, control of propulsion machinery from the navigation bridge, communications, alarm systems, safety systems, special requirements for machinery, boiler and electrical installations and automatic control and alarm systems.

For offshore units, the IMO MODU Code[2] includes additional requirements for unmanned machinery spaces including: fire prevention, fire detection, firefighting, protection against flooding—bilge water level detection, bridge control of propulsion machinery, communications, alarm systems, safety systems, machinery, boiler and electrical installations.

Ships with modern automation systems such as remote control, alarm and monitoring systems make it possible to operate the engine room unmanned, at least part of the time. During day working this allows the engineers to execute planned maintenance and repairs or replace defective parts.

Additionally, an unmanned engine room has an influence on ship minimum safe manning and results in lower crew cost for the owner.

Classification societies may assign an additional class notation to the main class notation indicating that the engine room can be unmanned for a certain period when all applicable rules for unmanned class notation are fulfilled. Major classification societies therefore issue their own rules that are more detailed than the SOLAS Convention. Different Classification Societies have different class notations for ships with unattended machinery space.

Lloyds Register additional class notation for unattended machinery space operation is **UMS**. American Bureau of Shipping additional class notation for ships where the propulsion machinery space can be periodically unmanned and machinery can be controlled primarily from the navigation bridge is **ACCU**. DNVGL classification society assigns additional class notation **E0** for ships with periodically unattended machinery spaces where machinery, alarm and automation arrangements assure vessel safety that is equivalent to having the machinery spaces attended.

---

2. MODU Code—IMO Code for the Construction and Equipment of Mobile Offshore Drilling Units.

Chapter 4

IMO regulations

Chapter outline

Abstract

This chapter outlines statutory requirements for ships control and monitoring systems published by the International Maritime Organization (IMO) in the Safety of Life at Sea Convention (SOLAS Convention) and for offshore units in the Code for the Construction and Equipment Mobile Offshore Drilling Units (MODU Code).

4.1 SOLAS regulations for ship's process control and monitoring

4.1.1 SOLAS regulations for 'MANNED' machinery spaces on ships

Essential regulations for 'MANNED' machinery spaces control and monitoring are included in the SOLAS Convention, Chapter II-1, Part C Machinery Installations. The main regulations describing requirements for 'MANNED' machinery spaces are described below:

    Regulation 31 Machinery Controls

– All systems essential for the propulsion, control and safety of the ship shall be independent and failure of one system shall not influence other system;

- The speed, direction of thrust and the pitch (if applicable) shall be fully controllable from the bridge;
- The main propulsion machinery shall be provided with an emergency stopping device on the navigation bridge;
- Propulsion machinery orders from navigation bridge shall be indicated in engine control room and locally;
- Remote control of the machinery shall be possible only from one location at a time;
- It shall be possible to control the propulsion machinery locally in the case of failure any part of the remote control system;
- In case of remote control system failure an alarm signal shall be given;
- An alarm shall be initiated on navigation bridge and in engine control room to indicate low starting air pressure.

Regulation 37 Communication between Navigation Bridge and Machinery Space

- Two independent means shall be provided for communicating orders from the bridge to engine control room and engine local control space;
- One of above communication means shall be engine room telegraph.

Regulation 38 Engineer's Alarm

- Engineer's alarm clearly audible in engineer's accommodations shall be provided, and is to be operated from engine control room and engine local control space.

The exact wording of the regulations can be found in the up-to-date and ratified by member states Consolidated Edition of the SOLAS Convention.

4.1.2 SOLAS regulations for periodically 'UNATTENDED' machinery spaces on ships

Essential regulatory regulations for periodically 'UNATTENDED' machinery spaces control and monitoring are included in SOLAS Convention, Chapter II-1, Part E Additional Requirements for Periodically Unattended Machinery Spaces. The main regulations describing requirements for periodically 'UNATTENDED' machinery spaces are described below:

Regulation 46 General

- Safety of the ship in all sailing conditions shall be equivalent to a ship having the machinery spaces manned.

Regulation 47 Fire Precautions

- Ships shall be equipped with fire alarm systems giving alarms at early stage in case of fire in boiler air supply casings and exhaust and in scavenging air of propulsion machinery,

Regulation 48 Protection against Flooding

– Accumulation of liquids in bilge wells shall be monitored;
– Automatic bilge pumps running time and starts frequency shall be monitored;
– Controls of sea inlet, water discharge or bilge injection system shall be installed in place to allow adequate time for operation in case of influx of water to the space.

Regulation 49 Control of propulsion machinery from the navigation bridge

– Speed, direction of thrust and pitch of the propeller (if applicable) shall be controllable from the bridge;
– Propulsion remote control shall be performed by a single control device;
– Propulsion machinery shall be protected from overload;
– Main propulsion machinery shall be provided with emergency stop on the bridge and stopping device shall be independent from the control system;
– Propulsion machinery orders from the bridge shall be indicated in engine control room or engine local control station
– Propulsion machinery control shall be possible from one location at a time;
– It shall be possible to control all propulsion machinery from local position;
– In case of propulsion machinery failure, alarm shall be initiated;
– Bridge indicators shall be fitted for propeller speed, direction of rotation and propeller pitch (if applicable).

Regulation 50 Communication

– Voice communications shall be provided between engine control room, bridge and engineer's offices and accommodation.

Regulation 51 Alarm System

– Alarm system shall be provided to indicate machinery abnormal machinery condition and visually indicate each alarm separately;
– Alarms shall be transferred to engineer's public rooms and their cabins;
– Alarms requiring action shall be transferred to the bridge;
– Alarm system shall be designed on fail-to-safe principle (as far as practicable) and power shall be redundantly supplied;
– Alarm system shall be able to simultaneously indicate more than one fault;
– Alarms shall be maintained until accepted.

Regulation 52 Safety Systems

– Safety system shall be provided to ensure that serious malfunctions in machinery or boiler operations that present immediate danger are eliminated by shutting down faulty equipment automatically;
– Shutdown of the propulsion system shall not be automatically activated except in cases which would lead to serious damage, complete breakdown or explosion.

Regulation 53 Special Requirements for Machinery, Boiler and Electrical Installations

- In case of loss of one generator during parallel operation of two or more generators, provisions shall be made to not overload the remaining generator(s) by load shedding;
- Equipment needed for main propulsion operation shall be provided with automatic arrangements.

Exact wording of the regulations can be found in up-to-date and ratified by member states Consolidated Edition of the SOLAS Convention.

4.2 MODU code regulations for mobile offshore units machinery spaces

4.2.1 MODU code regulations for 'MANNED' machinery spaces on mobile offshore units

Essential regulatory regulations for 'MANNED' machinery spaces control and monitoring are included in the MODU Code:

Chapter 4 Machinery Installations for all types of Units. The main control and monitoring regulations describing requirements for 'MANNED' machinery spaces are described below:

- Central ballast control station should be provided with the following functions—ballast pump control and monitoring system, ballast valve control and monitoring system, tanks level and draught indicating system, heel and trim indicators, electrical power and hydraulic/pneumatic pressure indicating system;
- Internal communication—shall work independently from unit's main source of electrical power;
- Protection against flooding—all seawater inlets and discharges below the assigned load line should be provided with valves operable from a position outside the space; all control systems and indicators should be operable in the event of unit main power failure.

Chapter 5 Electrical Installations for all types of Units. The main control and monitoring regulations describing requirements for 'MANNED' machinery spaces are described below:

- Main source of electrical power—when electrical power is supplied from more than one generator in parallel operation, provision should be made to ensure that generators are not overloaded if one is lost;
- Emergency source of power—emergency generator, the transitional source of emergency power, and emergency switchboard shall be located above the worst damaged waterline and available power shall be sufficient to supply all services that are essential for safety in emergency, emergency generator shall be equipped with two independent means of starting and shall be

started and connected to emergency load as quickly as possible but not later than 45 seconds from main generator black out;

– Alarms and communications—each unit shall be provided with a general alarm system giving alarm in case of general emergency, toxic gas (hydrogen sulphide), combustible gas, fire alarm and abandon unit signals;

– Public address system—shall be provided with possibility to make announcements from Emergency response centre, navigation bridge, engine control room and ballast control station.

Chapter 7 Machinery and Electrical Installations for Self-propelled Units. The main control and monitoring regulations describing requirements for 'MANNED' machinery spaces are described below:

– Machinery controls—all control systems essential for the propulsion, control and safety of the unit should be independent or designed so that failure of one system does not degrade the performance of another system;

– Remote control of propulsion machinery from navigation bridge—the speed, direction of thrust and pitch (if applicable) should be fully controllable, remote control should be provided for each independent propeller, emergency stopping device shall be provided, remote control should be indicated in engine control room, remote control of the propulsion machinery should be possible from only one station at a time, an alarm should be initiated at the navigation bridge and in machinery control room to indicate main engine low starting air pressure;

– Communication between the navigating bridge and the engine room—two independent means of communication shall be provided;

– Engineer's alarm—alarm from engine control room or manoeuvring platform should be clearly audible in engineers' accommodation.

4.2.2   MODU code regulations for 'UNMANNED' machinery spaces on mobile offshore units

Chapter 8 of the MODU Code contains additional essential regulations for 'UNMANNED' machinery spaces control and monitoring. The primary principle of the regulations is that safety of unattended machinery space operation shall be equivalent to manned machinery spaces. These additional regulations include:

– Fire prevention—protection of oil fuel and lubrication oil pipes, preventing from overflow spillages of daily service oil fuel tanks filled automatically, providing high temperature alarm for heated daily service oil fuel tanks or setting tanks;

– Fire detection—the detection system should initiate audible and visual alarms clearly distinguishable from other systems in sufficient locations to ensure they are heard and observed by the crew, the system shall be fed from main and emergency sources of power and separate feeders with automatic change over switch;

- Fire fighting—the fire main system shall be permanently pressurised, or main fire pumps starting arrangements shall be provided on navigation bridge and other normally manned control station;
- Bilge water detection—high bilge water level with audible and visual shall be provided, indication in manned location that influx of liquid is greater than automatically started bilge pumps capability;
- Bridge control of propulsion machinery—speed, direction of thrust and pitch (if applicable) should be fully controllable from the navigation bridge, the main propulsion machinery shall be provided with emergency stop on navigation bridge, remote control of the propulsion machinery shall be possible from only one location at a time, it should be possible to control all machinery essential for propulsion and manoeuvring from a local position, alarm shall be given on the navigation bridge and at the main machinery control station in case of remote automatic control failure, indicators shall be fitted on navigation bridge for propeller speed and direction of rotation or pitch position, an alarm shall be provided to indicate low starting air pressure;
- Communication—vocal communication shall be provided between the main machinery control station, navigation bridge, engineer offices, accommodation and central ballast control station (if applicable);
- Alarm system—an alarm system shall be provided in main machinery control station giving audible and visual indications on any fault requiring attention, audible and visual alarm shall be activated in normally manned control station, in the engineer's accommodation, offices and for the officer on watch (in marine mode), system shall be continuously powered with changeover to stand-by supply and alarm whenever there is failure of normal power,
- Automatic control and alarm systems—the services needed for the operation of the main propulsion and machinery shall be automatically controlled, a monitoring and alarm system shall indicate all important pressures, temperatures and fluid levels necessary for propulsion;
- Safety systems—shall be provided to ensure that any serious malfunction in machinery or boiler operations which presents an immediate danger shall initiate automatic shutdown of that part of the system and an alarm shall be given.

Chapter 5

Classification societies rule requirements

Chapter outline

Abstract

This chapter outlines rule requirements for control and monitoring systems published by two out of the twelve class societies belonging to the International Association Classification Societies (IACS)[1] i.e. American Bureau of Shipping (ABS) and Det Norske Veritas Germanischer Lloyd (DNVGL). The Classification Societies rules contain obligatory requirements to assign ship class and additional class notation after control and monitoring documentation approval, installation and survey on board.

1. IACS web page http://www.iacs.org.uk/.

Ship and Mobile Offshore Unit Automation. https://doi.org/10.1016/B978-0-12-818723-4.00005-3
© 2019 Elsevier Inc. All rights reserved.

5.1 American Bureau of Shipping (ABS) rules for building and classing steel vessels[2]

ABS Rules for Building and Classing Steel Vessels in Part 4, Chapter 9 Vessels Systems and Machinery includes requirements for control and monitoring systems:

- Essential features requirements—Systems performing different functions are to be as far as practicable independent so that single failure in one will not influence the other system. Local manual controls shall be fitted to enable safe operation of the system during maintenance and commissioning. Remote controls shall be arranged to assure the same degree of safety and operations as for local controls. Control, monitoring and safety systems shall be supplied from monitored separate circuits.
- Monitoring and alarm systems—Visual and audible alarms shall not effect operation of each other. The system shall have self-monitoring features and built-in alarm testing.
- Safety System—Abnormal system parameters causing safety action shall be distinguishable. Safety system activation shall be indicated on the navigation bridge, at the centralised control station and local manual control positions. Remote propulsion control system shall provide an individual alarm for automatic shutdown and slowdown and it should be possible to override safety system functions.
- Remote Propulsion Control System—Remote System shall be as effective as local control and shall be provided with sufficient instrumentation including speed and propeller direction.
- Remote Propulsion Control on Navigation Bridge—An emergency stop device for the propulsion machinery independent from the propulsion control system shall be provided on the navigation bridge. Alarm shall be provided to indicate low level starting medium energy and the control system shall limit number of consecutive automatic starts of the propulsion machinery to safeguard sufficient capacity for local manual starting.

Additionally, the ABS rules include a table with detailed requirements for Instrumentation and Controllers on Remote Propulsion Control Stations.

Special requirements are listed for Computer-based Systems:

- System Requirements—Systems shall be provided with effective security arrangements with restoration facility after power failure and self-monitoring functions. Control, monitoring and safety systems shall be arranged so that single failure or computer malfunction will not affect more than one of these system functions. Computer response time for propulsion related applications shall not exceed two (2) seconds. Emergency stops shall be hard-wired and be independent from computer-based systems.

---

2. Based on ABS web page https://ww2.eagle.org/en/rules-and-resources/rules-and-guides.html#/content/dam/eagle/rules-and-guides/current/conventional_ocean_service/2_steel_vessel_rules_2018.

- System Configuration—Systems shall be assigned to appropriate system category I, II or III according to possible damage they could cause for the controlled system. Based on the assigned category, the system software shall follow Software Development Life Cycle (SDLC) concept and tests.
- Hardware—System hardware shall ensure easy access to exchangeable parts for repairs and maintenance, User interfaces shall avoid inadvertent operator errors and shall be provided with security arrangements to limit access to authorised operator only. Alarms shall be displayed on Visual Display Unit(s) (VDU) regardless of the mode the computer or VDU is in. Alarms shall be clearly distinguishable from other information and shall be visually and audibly presented. Hardware system/elements shall be type tested according to system category I, II or III.

5.1.1   Automatic Centralised Control—ACC Notation

ABS assigns the ACC notation for ships with manned propulsion machinery spaces equipped with centralised control station to control and monitor the propulsion and auxiliary machinery. Remote controls include remote propulsion control, putting on-line and starting/stopping of the standby generators, and start, stop and transfer of auxiliaries necessary for the operation of propulsion and power generation.

ABS Rules for Building and Classing Steel Vessels in Part 4, Chapter 2 Vessels Systems and Machinery contains requirements for control and monitoring from Centralised Control Stations that include:

- Operator stations—at least two computers, including keyboards and monitors;
- Engineer's Alarm—where alarms are not acknowledged in a pre-set time period this system activates the alarm in the senior engineer's accommodation and specified public spaces;
- Propulsion Machinery Space—Low level of fuel in oil setting and service tanks must be alarmed, high level of fuel in oil drain tank receiving fuel from drip pans and spill trays shall be alarmed. Heated fuel oil tanks are to be equipped with fuel oil high temperature in the settling and service tank alarms, low flow alarm at heater outlets, high temperature alarm for fluid heating medium. Bilge level monitoring is to be provided with two independent systems detecting excessive rise of bilge water in bilges or bilge wells, bilge pump alarm shall be initiated when the pump is operating more frequently than usually. Fire detection and alarm system is to be provided in propulsion machinery space, main fire pumps to be capable of remote start from navigation bridge.

Additionally, the rules include tables with detailed requirements for Instrumentation and Controllers in Centralised Control Stations—All Propulsion and Auxiliary Machinery. The table contains requirements for propulsion control and monitoring, electric power generating plant, high voltage rotating machines, fuel oil system, stern tube lube oil, thermal oil boiler, heater, incinerator, etc.

The propulsion machinery space table shows if alarm, display or control is to be applied.

5.1.2 Automatic Centralised Control Unmanned—ACCU notation

ABS assigns the ACCU Notation for ships with periodically unmanned machinery space allowing control of propulsion machinery from the navigation bridge. The extent of automation, monitoring and remote control depends on the duration of unattended operation. Ships with ACCU notation shall be fitted with a remote propulsion control station on the navigation bridge, a centralised control station, plus a monitoring station in the senior engineer's living quarters and fire fighting station.

ABS Rules for Building and Classing Steel Vessels in Part 4, Chapter 2 Vessels Systems and Machinery includes additional requirements for:

- Propulsion and Auxiliary Machinery—The auxiliaries that are essential for propulsion and manoeuvring are to be automatically started, so that power is automatically restored following blackout.
- Monitoring Station in the Senior Engineer's Living Quarters—shall be provided with an alarm for fire in the propulsion machinery space, an alarm for high bilge water level in the propulsion machinery space, and a general/ summary machinery alarm. Audible alarms are to be silenced only at the centralised control station.
- Fire Fighting Station—shall be provided and must be located outside the propulsion machinery space. Manual controls shall include: shutdown of ventilation fans serving the machinery spaces, shutdown of fuel oil, lubrication oil and thermal oil system pumps, shutdown of draft fans for boilers, incinerators, auxiliary blowers for propulsion diesel engines, closing of propulsion machinery space fuel oil tanks suction valves, closing of propulsion machinery space skylights, openings in funnels, ventilator dampers, closing of propulsion machinery space watertight and fire-resistant doors, starting a fire pump from a location outside of the propulsion machinery space;
- Fire Detection and Alarm System—a fixed system shall be provided and a fire control panel is to be located on the navigation bridge and in the fire-fighting station, system to be provided with fire alarm call points in the Centralised Control Station, passageways leading to the propulsion machinery spaces and on the navigation bridge.

Additionally, the ACCU rules contain tables with requirements for instrumentation and controllers in the Centralised Control Station for:

- Slow, Medium and High Speed diesel engines—detailing requirements for fuel oil, lubrication oil, turbocharger, piston cooling, sea water cooling, cylinder fresh water cooling, compressed air, scavenge air, exhaust gas, fuel valve coolant, engine and power.

- Propulsion steam turbines—detailing requirements for lubrication oil, lubrication oil cooling medium, sea water, steam, condensate and turbine;
- Propulsion gas turbines—detailing requirements for fuel oil, lube oil, cooling medium, starting, combustion, exhaust gas and turbine;
- Electric propulsion—detailing requirements for propulsion generator, propulsion motor AC or DC, propulsion SCR and transformer;
- Generator prime mover for electric propulsion—detailing requirements for fuel oil, lube oil, sea cooling water, cylinder fresh water cooling, compressed air, exhaust gas, turbocharger, engine, power supply, and for gas turbines—fuel oil, lubrication oil, cooling medium, starting, combustion, exhaust gas, turbine and power supply;
- Propulsion Boiler—detailing requirements for Feed water, Boiler drum, Steam, Air, Fuel oil, Burner and Power;
- Auxiliary Boiler—detailing requirements for Feed water, Boiler drum, Steam, Air, Fuel oil, Burner and Power;
- Auxiliary turbines and diesel engines—detailing requirements for diesel engine lube oil, cooling medium, fuel oil, starting medium, exhaust, speed and for steam turbine lube oil, lube oil cooling medium, sea water, steam, condensate, rotor and for gas turbine lube oil, cooling medium, fuel oil, exhaust gas, combustion, starting and turbine.

5.1.3 Automatic Bridge Centralised Control Unmanned—ABCU notation

ABS assigns the ABCU Notation for ships with automation and remote monitoring and control systems which enable the propulsion machinery space to be periodically unattended similar to ACCU class and the propulsion control to be effected primarily from the navigation bridge.

5.2 American Bureau of Shipping (ABS) rules for building and classing mobile offshore drilling units

ABS Requirements[3] for MODU Control and Monitoring Systems are similar to those for Steel Vessels included in ABS Rules for Building and Classing Steel Vessels, Part 4, Vessels Systems and Machinery, with some additional requirements as follows:

- Interior Communication Systems—At least two independent systems shall be provided for orders communicating from navigation bridge to machinery control room, one of these is Engine Order Telegraph with visual indication of orders and responses. Voice communication shall be provided between the main propulsion control station and local control positions for

---

3. Based on ABS web page https://ww2.eagle.org/en/rules-and-resources/rules-and-guides.html#/content/dam/eagle/rules-and-guides/current/offshore/6_mobile_offshore_drilling_units_2018.

main propulsion. Voice communication shall be provided between navigation bridge, main propulsion control station and steering gear compartment.
Voice communication shall be available between all locations where action
may be necessary in case of an emergency. Internal communication power
supply shall be independent of control, monitoring and alarm system supply.
- Public Address System—A loud speaker system shall enable the broadcast
  of messages from the navigation bridge and emergency control stations, and
  such messages shall be audible in all parts of the unit. Public address system
  shall be supplied from an emergency source of power. When Public address
  system is combined with General Alarm System then a single failure shall not
  cause the loss of both systems, and effect of single failure shall be minimized.
- General Emergency Alarm System—System shall be supplied from main and
  emergency supply and shall be capable of sounding the general emergency
  alarm signal, fire alarm signal and abandon unit signal. In case of loss of
  power in one of the feeders supplying the general alarm system, an alarm shall
  be provided in a normally manned control station. The system shall be capable
  of operation from the navigation bridge and emergency control stations.
- Ventilation Emergency Stop System—System shall be provided to stop ventilation fan motors in case of fire for machinery spaces, propulsion machinery
  spaces, accommodation spaces, control stations, service and other spaces.
- Emergency Shutdown Facilities—Emergency shutdown system shall be
  arranged to disconnect selectively or simultaneously all affected electrical
  equipment and devices according to guidance from a Functional Design
  Basis Document (FDS) and Fire and Gas Cause and Effect (C&E) Chart.

5.3 DNVGL rules for classification of ships

DNVGL Rules for Classification Ships[4] in Part 4, Chapter 9 Control and
Monitoring Systems includes requirements as follows:

- Design principles—Instruments belonging to separate essential processes
  shall be independent. An integrated system shall be arranged with sufficient
  redundancy to prevent loss of essential functions upon a single failure.
- Response to failure—Essential systems shall detect failures that may cause
  system errors and detected errors shall initiate alarms. Redundant systems
  shall have self-diagnostic properties and upon fire shall transfer data to a
  stand by unit.
- System design—Automatic control shall keep process variables within
  the limits during normal operational conditions. Remote control user shall
  receive continuous information on control effects. Control shall not be
  transferred before being acknowledged and during the control transferring

---

4. Based on DNVGL web page https://rules.dnvgl.com/ServiceDocuments/dnvgl/#!/industry/1/
Maritime/1/DNV%20GL%20rules%20for%20classification:%20Ships%20(RU-SHIP).

process parameters shall not be changed significantly. Protective safety functions shall be forewarned by appropriate alarms and it shall be possible to trace the cause of the safety system action. Alarms shall be announced by visual and audible signals, responsibility for alarms shall not be transferred before acknowledgment by the receiving location. It should be possible to delay alarms to prevent incorrect alarms due to normal transient conditions. Indicators shall be installed for essential control functions to allow systems safe operation.

- General requirements—Two independent power supplies shall be provided for essential control, monitoring and safety systems. Important services for redundant control, monitoring and safety systems shall be supplied by independent power supplies.

Special requirements are listed for Computer-based Systems:

- Computer based systems response time shall be less than $0.1\,s$ for data sampling, $1\,s$ for other indications, $2\,s$ for alarm presentations and update of screen views, and $5\,s$ for fully updated screen views including start of new application.
- System access—Access to computer system set up and configuration shall be protected by password to avoid unauthorised modifications.
- Requirements for computer-based system software—control system networks and data communication links are detailed in several specific chapters of the rules.

In addition, DNVGL rules include requirements for instrumentation environmental conditions. Instrumentation parameters for temperature are divided by classes A, B, C, D; for humidity by classes A, B; for vibration by classes A, B, C and for Electromagnetic Compatibility by classes A and B. For each class there are specified different test limits. Instrumentation is tested also for resistance to salt contamination, oil contamination and inclination, if applicable. Separate tables contain requirements for electromagnetic compatibility, minimum immunity and maximum emission parameters.

Additionally in Part 4 System Components Rules, there are included control and monitoring tables with detailed requirements for: control and monitoring of propulsion engines, control and monitoring of auxiliary engines, engines for emergency generators and gas-fueled engines, monitoring of shafting, gear transmissions, clutches and elastic couplings for single diesel engine propulsion plants, control and monitoring of water jets and thrusters, compressors and exhaust gas cleaning systems for the reduction of Nitrogen Oxide air pollution (NOx), monitoring and recording of exhaust gas cleaning systems for the reduction of Sulfur Oxide air pollution (SOx), plus monitoring of electrically heated pressure vessels, oil fired auxiliary boilers, oil-fired water heaters, thermal-oil heaters and oil burners. In System Components Rules there are also included lists of required alarms and monitoring parameters for miscellaneous electrical equipment, and control and monitoring of steering gear.

5.3.1 Additional class notation ECO

Where the main propulsion and associated machinery, including sources of main electrical supply, are provided with various degrees of automatic or remote control and are under continuous manual supervision from a control room, and the arrangements and controls and monitoring system meet SOLAS II-1/31.1 then the system can be given the ECO notation by DNVGL. The required scope of control and monitoring equipment is given in the Part 4 System Components tables listed in the subchapter above.

5.3.2 Additional class notation E0

Additional class notation E0 for periodically unattended machinery spaces is assigned for ships when the main propulsion control system meets appropriate rules requirements, SOLAS Convention SOLAS II-1/E requirements, and alarms are displayed on the bridge and in engineer's accommodation. Control and monitoring systems shall ensure the safety of the ship in all sailing and maneuvering conditions is similar to that of a ship with the machinery spaces manned.

To assign E0 Class notation the ship's control and monitoring systems shall not only fulfil DNVGL Rules for Classification Ships Part 4, Chapter 9 Control and Monitoring Systems but additionally shall fulfil Part 6, Chapter 2 Additional Class Notations which gives requirements for:

- Safety actions: Safety shutdowns for propulsion shall be executed automatically. Alarm of activation shut down shall be independent from the main alarm system. It should be possible to override automatic shutdowns not protecting the propulsion engine from immediate break down. Scope of bridge alarms: Alarms shall include automatic shut-down of main boiler, propulsion machinery, request for manual shutdown of propulsion machinery, failure of propulsion machinery remote control, steering control, bridge alarm supply and low starting air pressure for propulsion engines.
- Electric power supply: Arrangements shall be provided to protect generating sets overload.
- Fire safety: Oil pressure pipelines shall be shielded to avoid oil spray or leakages to hot machinery surfaces or into machinery intakes.

In Part 6 Additional Class Notations, the DNVGL rules include additional control and monitoring tables for E0 notation with detailed requirements for: control and monitoring of propulsion engines and propulsion turbines, main steam and feed water installations, shafting, propellers, gears, clutch and elastic couplings, auxiliary engines and auxiliary turbines, auxiliary boilers, and electrical power plant, together with monitoring of miscellaneous objects and tanks.

5.3.3   Supplementary requirements for drilling units

DNVGL Offshore Standard DNVGL-OS-D202 Automation, Safety and Telecommunication Systems, Section 6 describes supplementary requirements for mobile offshore units.[5] This standard requires the following: Automation and safety system components intended to be alive after an accident shall be certified for Ex Zone 2 or shall be safe by location. User interface back-up means of operation shall be provided for the most important functions and alarm indications related to emergency, gas detection and activating of fire protection devices. An Alarm Philosophy shall be developed for various alarm conditions.

5.3.4   Supplementary requirements for production and storage units

DNVGL Offshore Standard DNVGL-OS-D202 Automation, Safety and Telecommunication Systems,[6] Section 7 describes supplementary requirements for floating production units—Important shutdown devices shall only be reset locally, automation and safety system components intended to be alive after an accident shall be certified for Ex Zone 2 or shall be safe by location, F&G nodes shall be redundant and signals between F&G and ESD nodes shall be fail safe. An Alarm Philosophy shall be developed for various alarm conditions. User interface back-up means of operation shall be provided for most important functions and alarm indications related to emergency, gas detection, and activating of fire protection devices.

DNVGL Offshore Standard DNVGL-OS-D202 Automation, Safety and Telecommunication Systems, Section 8 describes supplementary requirements for storage units and requires that important shutdown devices shall only be reset locally.

---

5. Based on DNVGL web page https://rules.dnvgl.com/ServiceDocuments/dnvgl/#!/industry/2/Oil%20and%20Gas/14/DNV%20GL%20standards%20(ST).
6. Based on DNVGL web page https://rules.dnvgl.com/ServiceDocuments/dnvgl/#!/industry/2/Oil%20and%20Gas/14/DNV%20GL%20standards%20(ST).

Chapter 6

International codes and standards

Chapter outline

Abstract

This chapter outlines international regulations applicable to control and monitoring systems that are often referenced in contract documents between owners and shipbuilders or subcontractors. Many of the International Electro-technical Commission (IEC) Standards, Electrical and Electronics Engineers (IEEE) Standards, are often referenced in IMO Conventions and Codes and are thus becoming obligatory for vessel owners. European Union Directives are also obligatory for vessels flagged in EU states. This chapter additionally includes standards and regulations that often need to be followed in offshore projects, e.g. American Petroleum Institute[1] (API) Recommended Practices and Standards, Norwegian Maritime Authority (NMA) and NORSOK Standards and regulations.

6.1 International Electro-technical Commission (IEC) standards

The International Electrical Congress held on September 15th, 1904 in St. Louis USA decided to establish an International Electro-technical Commission[2] (IEC). The first office of the IEC was set up June 1906 in London, and since that time the IEC has been serving worldwide industries, including equipment and installations on ships and offshore units.

---

1. API webpage http://www.api.org/.
2. IEC web page http://www.iec.ch/standardsdev/publications/is.htm.

Ship and Mobile Offshore Unit Automation. https://doi.org/10.1016/B978-0-12-818723-4.00006-5

The International Electro-technical Committee publicises consensus based global standards for electrical design, installation, testing control and monitoring systems, equipment and instrumentation. IEC Standards are developed by consensus and are approved by means of a National Committees voting process. All IEC Standards are available for purchase in IEC Webstore[3]. These standards are often included within vessel/offshore unit project technical specifications as additional contractual requirements. For example:

- For ships—IEC 60092 series Electrical Installations in Ships, especially Part 504: Automation, Control and Instrumentation standard includes specific requirements for fire protection control installations, machinery control and alarm installations, automatic starting installations for electrical motor-driven auxiliaries, machinery safety systems, and computer-based systems. Additional requirements are given for periodically unattended machinery spaces or for reduced attendance with fire precautions, protection against flooding, control of propulsion machinery, alarm and safety systems and special requirements for machinery, boiler and electrical installations.
- For offshore units—IEC 61892-1 to 7 Mobile and Fixed Offshore Units—Electrical Installations contains requirements for control and monitoring systems, equipment and installation.
- For ships and offshore units—IEC 60533, Electrical and Electronic Installations in Ships—Electromagnetic Compatibility standard specifies minimum requirements for emission, immunity and performance criteria regarding electromagnetic compatibility of electrical and electronic equipment on ships and offshore units.
- For installation of control and monitoring systems and equipment on ships and on offshore units—IEC 60079-0 to 19 contains requirements for control and monitoring systems, equipment and installation in explosive atmospheres.

The following standards are often used for tests and certification of control and monitoring equipment:

- IEC 60068-2, Environmental Testing—Part 2 for Cold, Dry heat, Vibration, Damp Heat, Salt Mist Tests,
- IEC 61000-4 Electromagnetic compatibility (EMC)—Part 4 contains requirements for electrostatic discharge immunity, radiated, radio-frequency, electromagnetic field immunity, electrical fast transient/burst immunity, surge immunity, immunity to conducted disturbances, induced by radio-frequency fields, voltage dips, short interruptions and voltage variations immunity tests.

The following standards are often used for test and certification of cables for ships/offshore units control and monitoring systems:

- IEC 60331 for fire resistant cables,
- IEC 60332-3 for fire retardant cables,

---

3. IEC Webstore web page https://webstore.iec.ch/?ref=menu.

- IEC 61034 for low smoke cables,
- IEC 60754-1 for halogen free cables,
- IEC 60092-352 for cables insulation material,
- IEC 60092-359 for cables sheathing material.

International Electro-technical Committee (IEC) Standards titles often referenced in ship and mobile offshore Unit automation systems are listed in Appendix 4.

6.2   Institute of Electrical and Electronics Engineers (IEEE)

Institute of Electrical and Electronics Engineers[4] is a UK professional body with the objective of technical advancement of electrical and electronic engineering, telecommunications and computer engineering disciplines. IEEE professionals are working in thirty-nine different groups, for example: Computer Society, Control Systems Society, Electromagnetic Compatibility Society, Information Theory Society, Instrumentation & Measurement Society, Power Electronics Society, Robotics and Automation Society, Signal Processing Society etc.

Among other publications, the IEE publishes standards for electrical and electronic equipment and systems onboard ships. For control and monitoring systems of ships and offshore units the technical specifications often refer to the IEEE 802 Series Standards for Information Technology including Ethernet, LAN, Wi-Fi, MAN, WAN, Token Ring, Bluetooth. Other examples of IEEE Standards sometimes quoted in marine and offshore project technical specifications are: IEEE 1284 Parallel Interface for Personal Computers, IEE1394 Serial Bus, IEEE 1003 Portable Operation Systems, IEEE 1174 Serial Interface for programmable Instrumentation, and IEEE 45 Recommended Practice for Electric Installations on Shipboard.

6.3   European Union directives

The European Union issues legal acts—Directives[5], which require Member States to achieve the outlined target results, and compliance is obligatory for European Union States. Directives are numbered sequentially by each year and can be found online on the EUR-lex webpage. Some Directives are applicable to equipment and control and monitoring systems installed on ships and offshore units, including:

- Directive on Marine Equipment (MED) with the objective to enhance safety at sea and prevent marine pollution by using certified equipment on EU Member state vessels falling with IMO Statutory Requirements. Marine equipment complying with this Directive is marked with the "wheel mark".

4. Based on webpage https://www.ieee.org/index.html.
5. Based on webpage http://eur-lex.europa.eu/homepage.html.

- Directive on Machinery with objective to improve the safety of machinery placed on the market. Machinery complying with this Directive is marked 'CE'.
- Directive on the harmonisation of the laws of the Member States relating to electromagnetic compatibility with aims to ensure compliance of electrical equipment with an adequate level of electromagnetic compatibility. Equipment shall be designed and manufactured to ensure that the electromagnetic disturbance generated does not exceed the level at which radio and telecommunication equipment ceases to operate as intended, and has a level of immunity to electromagnetic disturbance to be expected in its intended installation. Electrical equipment is tested according to requirements of harmonised standards adopted by European Union, published in EU Official Journal and afterwards equipment is marked 'CE'.
- Directive on the harmonisation of the laws of the Member States relating to equipment and protective systems, safety devices and controlling devices intended for use in potentially explosive atmospheres. According to this Directive the equipment is determined by groups and categories symbols. Electrical equipment complying with this Directive requirements is marked 'CE', and the specific marking of explosion protection 'Ex' followed by the symbol of the equipment group and category. In the UK this is referred to as ATEX compliance.

6.4 American Petroleum Institute (API) recommended practices and standards

Contractual obligations for offshore units in the oil and gas industry often include American Petroleum Institute[6] (API) Recommended Practices and Standards. These API publications are designed to assist industry professionals improve the efficiency and cost effectiveness of their operations and comply with valid legislative and regulatory requirements. Some of the electrical and control related publications of API include the following:

- Recommended Practice for Classification of Locations for Electrical Installations at Petroleum Facilities Classified as Class I, Division 1 and Division 2 (RP500)—including guidelines for determining electrical equipment selection and installation in different locations where there may be a risk of ignition due to the presence of flammable gas or vapour mixed with air.
- Recommended Practice for Classification of Locations for Electrical Installations at Petroleum Facilities Classified as Class I, Zone 0, Zone 1 and Zone 2 (RP505)—including guidelines for determining the degree and extent of Class I, Zone 0, Zone 1 and Zone 2 locations.

---

6. Based on webpage http://www.api.org/.

- Process Measurement Instrumentation (RP551)—provides procedures for measuring and control instruments installation.
- Process Control Systems (RP554)—including requirements for functions, functional specification development, process control system design

6.5 Coastal state rules for offshore units

6.5.1 Norwegian Maritime Authority (NMA)

Offshore installations on Norwegian coastal waters shall meet the requirements of the Norwegian Maritime Authority (NMA) and shall fulfil all requirements included in NMA 'Regulations relating to Maritime Electrical Installations'. These regulations apply to design, construction, operation and maintenance of maritime electrical installations and are focussed on fire and electrical safety, but also include rules relating to control and monitoring systems.

Obligatory requirements for control and monitoring systems on offshore units on Norwegian coastal waters are included in NMA 'Regulations for Mobile Offshore Units'. These regulations include requirements concerning precautionary measures against fire and explosion on offshore units and requirements for alarm and intercommunication systems.

6.5.2 NORSOK standards

The Norwegian petroleum industry developed NORSOK Standards[7] during the 1990s to ensure adequate safety, while at the same time aiming at standardisation and cost effectiveness for the offshore petroleum industry developments and operations. These standards are generally specified for offshore units working on the Norwegian shelf, and some are mandatory (e.g. Working Environment). The following NORSOK Standards are applicable for control and monitoring systems:

- I-001—Field Instrumentation—including functional and installation requirements for field instrumentation.
- I-002—Safety and Automation system (SAS)—including functional and system requirements for safety automation systems performing monitoring, logic control and safeguarding offshore installations.

---

7. Based on webpage https://www.standard.no/en/sectors/energi-og-klima/petroleum/norsok-standards/#.WpgHzmepXYU.

Chapter 7

Additional client requirements related to automation systems safety

Chapter outline

Abstract

This chapter describes some examples of additional requirements imposed by owners of the vessels and offshore units. These requirements often arise from previous experience or accidents which occurred on similar vessels or installations. Such additional requirements include development of Alarm Management Philosophies and or determining the Safety Integrity level (SIL) for onboard safety systems.

7.1 Alarm management philosophy

7.1.1 General

Control and Monitoring systems on vessels and offshore units, especially FPSOs, are very sophisticated and their operators receive enormous quantities of information and alarms. In the past poor operational awareness has caused many accidents because the operators of the Control and Monitoring systems did not respond with appropriate actions. Investigation of such situations frequently reveals that control panel graphics did not present the necessary process

Ship and Mobile Offshore Unit Automation. https://doi.org/10.1016/B978-0-12-818723-4.00007-7

overviews, and operator response was often compromised due to an excessive number of alarms occurring in emergency situations.

Having in mind such accidents exacerbated by the automation systems, owners of vessels and offshore units are often asked by their clients to provide an Alarm Management Philosophy for approval. An Alarm Management Philosophy is a document that defines the governance of alarms on the vessel or offshore unit.

Unit owners may obtain the Alarm Management Philosophy from the shipyard in the case of new builds and may also develop the philosophy by themselves, often with Consultant help. The Alarm Management Philosophy describes the main alarm functions with warning and logging of abnormal situations, the role of the operator, taking into account human limitations, use of alarm priorities and acceptance. Such documents contain a structured methodology of alarm management describing alarm justification, operator needs, performance targets and guidance.

An Alarm Management Philosophy is developed based on sub-philosophies for installed systems:

- Integrated Automation System and included Alarm and Monitoring System Philosophy
- Safety Management System and included Fire Alarm and Public Address/ General Alarm Philosophies
- Functional alarm system i.e. Dynamic Positioning System on DP controlled vessels, production, storage and offloading systems on FPSO vessels.

The Alarm and Monitoring System Philosophy for the Integrated Automation System is usually developed and implemented by the system producer.

7.1.2 Legislation, guidance and standards

The following legislation, guidance and standards may be followed in order to prepare the philosophy:

- IEC 62682 *Management of Alarm Systems for the Process Industries*. This standard specifies principles and processes for alarm management systems based on distributed control systems and computer-based Human-Machine Interface technology. International Electro-technical Commission (IEC) Standards are recognized worldwide and applied by industry. IEC has also published complementary standards IEC 61508 *Functional Safety— Electrical, Electronic & Programmable Electronic Safety-Related Systems* and IEC 61511 *Functional Safety—Safety Instrumented Systems for Process Industry Sector.*
- EU Directive 2012/18/EU. European Union Directives are obligatory in European Union countries. The Directive is recognized by the UK HSE. Directive 2012/18/EU takes into account best practices for monitoring and control to reduce the system failure and control associated risks.

- EEMUA 191—*Alarm Systems—A Guide to Design, Management and Procurement*. This Engineering Equipment and Material Users Association (EEMUA) guide emphasizes human factors and is often used in UK industry.
- ISA-18.2 *Management of Alarm Systems for the Process Industries*. The International Society of Automation (ISA) is an American association that sets the standards for automation and control systems used by industry. The standard is often implemented by industry in the USA.
- YA-711 *Principles for Alarm System Design* published by the Norwegian Petroleum Safety Authority (PSA). This covers basic principles and guidelines on alarm generation, structuring, prioritization, and presentation for offshore installations on the Norwegian Continental Shelf.

The above standards and guidelines are focusing on timely response of operator to alarms initiated by equipment malfunction, process deviation or abnormal condition to improve processes safety, reliability, quality and efficiency.

7.1.3   Alarm management philosophy general chapters[1]

General chapters of an Alarm Management Philosophy usually contain the following elements:

- Definitions describing basic, aggregated and key alarms, alarms and signals handling by their filtering, validation, suppression, shelving, overriding and prioritization;
- System purpose—alarms to be usable in all system and processes conditions, and able to warn the operator about abnormal situations that require timely assessment and corrective actions to maintain vessel/offshore unit safety and fitness for purpose;
- Alarm generation—alarms to require an operator response and incoming signals shall be filtered and validated;
- Alarm structuring—alarms to be grouped and alarm suppression functions shall be included in the system;
- Alarm prioritization—alarms to be prioritized to help the operator to decide sequence of his activities when several alarms occur at the same time and the prioritization shall take into account the available time for necessary corrective actions to be performed;
- Alarm presentation—alarms to be presented using different colors on processes displays and on permanently shown Key Alarm list, and to be accompanied with an audible signal when new alarm occurs;
- Alarm handling—alarms to be accepted to confirm that the message has been read and understood.

---

1. Based on Petroleum Safety Authority YA-711 Principles for alarm system design www.ptil.no/getfile.php/135975/.

7.1.4   Alarm management philosophy specific chapters

Specific chapters in the Alarm Management Philosophy will be provided to describe specific requirements for presentation of alarms depending on their importance, including the priority level from Emergency to High, Low etc. and the required information level for each system. Depending on alarm priority, the sampling frequency is set for particular systems.

7.1.4.1   Alarm and monitoring system philosophy of integrated automation system

Specific chapters are developed for each particular vessel or offshore unit and describe alarm handling philosophies for systems such as:

- Propulsion—Main Engine(s) and Thrusters,
- Power Generation—Engines and Generators,
- Cargo monitoring and control,
- Machinery auxiliary systems—compressors, purifiers, Heating, Ventilation, Air Conditioning (HVAC), steam generators, bilges etc.

The Alarm and Monitoring Philosophy should take into consideration operator awareness on the Wheelhouse/Bridge, in the Engine Control Room, in the Cargo Control Room and in Central Control Room(s) for hydrocarbon production vessels.

Additionally, the Alarm and Monitoring System philosophy will describe the engineer's Extension Alarm System, Personnel Fitness System, Hospital Call and Refrigerator alarm systems.

7.1.4.2   Alarm and monitoring system philosophy of safety management system

For the safety systems, specific chapters are developed for each vessel or offshore unit. These chapters describe alarm handling philosophies for:

- Fire Alarm (FA) or Fire & Gas (F&G)
- Public Address (PA)
- General Alarm (GA)

7.1.4.3   Alarm and monitoring system philosophy of functional alarm systems

For DP vessels and FPSO etc. specific chapters are developed for alarm systems, for operational function systems and may include:

- DP Alarm Systems Philosophies describing alarm handling philosophies for Dynamic Positioning systems;
- Production, storage and offloading unit Process Control system—Alarm Systems Philosophies describing alarm handling philosophies.

7.1.5   Alarm management philosophy implementation

The Alarm Management Philosophy is implemented by improvements in the Human Machine Interface (HMI), for example by designing enhanced graphics with easy Alarm Help function, "ask" based displays to support abnormal situations and use of audio and visual methods to indicate alarm priority. The objectives are achieved by special tools and facilities developed to filter alarms by area/priority, alarm logging and reporting and alarm analysis.

An example of tools implementing an Alarm Management Philosophy is provided for the Kongsberg Integrated Marine Automation System K-Chief 700:

– Alarm priority is assigned to each alarm using the window shown in Fig. 7.1.1:

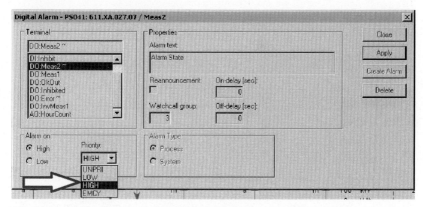

FIG. 7.1.1   Alarm priority assigning. *(Based on Kongsberg system print screens.)*

Alarms are marked on MIMIC diagrams and alarm list with colors: Green— UNPRI alarms, Yellow—LOW priority alarms, Red—HIGH priority alarms and Purple—EMCY alarms.

– Command control is configured for each Operator Station (OS) using the window shown in Fig. 7.1.2:

The primary function of Command Control configuration is to manage and distribute responsibility for systems control and alarms handling. A Command group under command by one Operator Station (OS) at the time is called 'an exclusive Command group' and a Command group under command of several Operator Stations is called a 'Shared command group'. The Operator Station receives events and alarms only from objects assigned to its command group. There are different types of Operator Stations groups:

– 'D' type—Operator Station takes a default group at start-up and when the OS is activated,

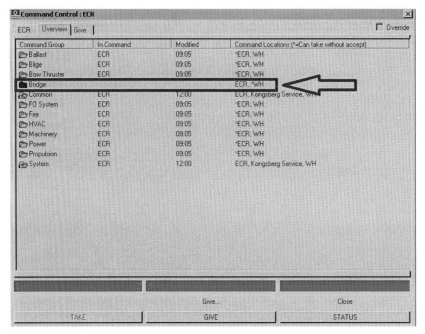

FIG. 7.1.2 Command control configuration. *(Based on Kongsberg system print screens.)*

- 'T' type—Operator Station can take a command group that is not under command of any other OS,
- 'R type—Operator Station can request the command, Give and Take

Some Command groups are 'S' type—under command of several OS groups simultaneously and 'DT' type—and can transfer control from its default location to another while the alarm responsibilities remains at the default location.

Monitoring Stations monitor process alarms without the possibility to control them which is available at Operator Stations.

Example of Command control details is presented in Fig. 7.1.3:

Another example of Alarm Management Philosophy implementation is the Extension Alarm System (EAS) which transfers alarm and monitoring system alarms and information to officers on duty in Wheelhouse, ECR, officer cabins, Public areas, CCR, and selected cabins. The Extension Alarm System contains extension alarm panels displaying activated alarms, identifying the on duty officers and location responsibility of alarms handle and enable operator response.

A typical Extension alarm system is presented in Fig. 7.1.4 Extension alarm system.

Alarms on the Extension Alarm System (EAS) are usually grouped. An example of such Alarm groups are presented in table in Fig. 7.1.5.

Groups command	Group type 'DT'	'S'	Wheelhouse 'D'	'T'	'R'	ACR 'D'	'T'	'R'	ECR 'D'	'T'	'R'	Serwer 'D'	'T'	'R'
Ballast	Yes	No	X	X										
Bilge	Yes	No			X	X	X				X			
Bow thrusters	Yes	No			X	X	X				X			
Bridge	No	No	X											
Common	No	Yes	X			X			X			X		
FiFi	Yes	No			X	X	X				X			
Fire	Yes	No			X	X	X				X			
FO system	Yes	No			X	X	X				X			
HVAC	Yes	No			X	X	X				X			
Machinery	Yes	No			X	X	X				X			
Power	Yes	No			X	X	X				X			
Propulsion	Yes	No			X	X	X				X			
System	Yes	Yes	X			X			X			X		
Thruster RCS	Yes	No			X	X	X				X			

FIG. 7.1.3 Command Control details. *(Based on Kongsberg system concept)*

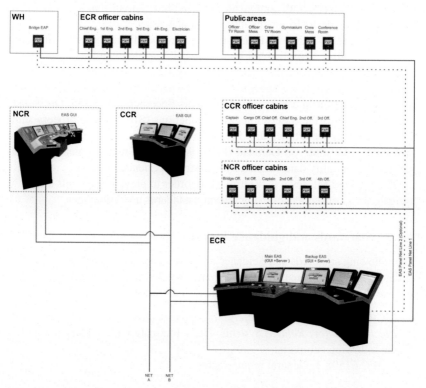

FIG. 7.1.4 Extension alarm system. *(Based on Kongsberg K-Chief 700 brochure.)*

Group number	Description	EAS panel text
1	Shutdown on engines and drives	SHUTDOWN
2	Load reduction alarm on engines and drives	LOAD RED.
3	Failure in switchboards or generators	POWER
4	Failure in engines and machinery equipment	MACHINERY
5	Failure in propulsion unit	PROPULSION
6	Failure in thrusters	BOW THRUSTERS
7	High level in bilge well	BILGE
8	Miscellaneous alarms	MISC.
9	System alarms	SYSTEM
10	Deadman	

FIG. 7.1.5   Example of alarm groups.

7.2   Safety integrity level

7.2.1   General

Vessels and offshore units operating in the oil and gas industry, especially FPSOs, are equipped with complicated automation systems that are intended to protect people, the environment and the physical asset.

Safety Integrity Level (SIL) is a description specifying the safety integrity level of the Safety Instrumented Functions (SIF), which is allocated to an appropriately designed safety instrumented system. Requirements for Safety Instrumented Systems are specified by the owner or the shipbuilder for a vessel or offshore unit in a Safety Requirements Specification (SRS). On vessels and offshore units the Safety Integrated Functions (SIF) include process shutdown, emergency shutdown and fire & gas functions. For these functions the Probability of Failure on Demand (PFD) is estimated using typical loop diagrams and reliability data verified by industry experience.

The required SIL level is determined in order to prevent accidents and/or to mitigate their consequences using relevant standards and guidelines.

7.2.2   Standards and guidance for SIL

The following standards and guidance are developed to standardize the process of determining performance standards, ensure consistency in the approach and achieve the required level of automation systems safety:

- IEC-61508 *Functional Safety of Electrical/Electronic/Programmable Electronic Safety-Related Systems*. These International Standards consist of the following sub-parts:
  - IEC 61508-1, General requirements;

- IEC 61508-2, Requirements for electrical/electronic/programmable electronic safety-related systems;
- IEC 61508-3, Software requirements;
- IEC 61508-4, Definitions and abbreviations;
- IEC 61508-5, Examples of methods for the determination of safety integrity levels;
- IEC 61508-6, Guidelines on the application of IEC 61508-2 and IEC 61508-3;
- IEC 61508-7, Overview of techniques and measures.
- IEC-61511 *Functional Safety—Safety Instrumented Systems for the Process Industry Sector*. These International Standards consist of the following sub-parts:
  - IEC-61511-1, Framework, definitions, system, hardware and application programming requirements;
  - IEC-61511-2, Guidelines for the application;
  - IEC-61511-3, Guidance for the determination of the required safety integrity levels.
- The following guidelines/frameworks provide best practice implementation of functional safety for automation systems:
  - ANSI/ISA-84, *Functional Safety: Safety Instrumented Systems for the Process Industry Sector* issued by the International Society of Automation (ISA). Often applied in USA.
  - CFR 1910.119, *Process Safety Management of Highly Hazardous Chemicals* issued by Occupational Safety and Health Administration (OSHA). Often applied in USA.
  - Recommended SIL requirements 070[2] – *Norwegian oil and gas application of IEC 61508 and IEC 61511 in the Norwegian Petroleum Industry.*

7.2.3 SIL determination methods

The SIL determination process is an analytical methodology consisting of the following procedures:

- Data gathering including P&ID, loop diagrams and other process specific data;
- Loops analysis;
- Common cause failures;
- Model development and execution;
- Results and conclusions.

Four SIL levels are defined in the relevant standards as shown in Fig. 7.2.1

---

2. Norwegian Oil and Gas Association Guideline 070 https://www.norskoljeoggass.no/en/working-conditions/retningslinjer/.

Safety integrity level (SIL)	Probability on failure on demand (PFD)	
1	0.1–0.01	10^{-1}–10^{-2}
2	0.01–0.001	10^{-2}–10^{-3}
3	0.001–0.0001	10^{-3}–10^{-4}
4	0.0001–0.00001	10^{-4}–10^{-5}

FIG. 7.2.1 Safety Integrity Levels (SIL).

The above SIL levels are defined for a low demand mode of operation where the frequency of demand for operation on safety related system is not greater than one per year and not greater than twice the proof test.

Safety Instrumented System (SIS) with SIL 1 rating reduce risk by a factor of 10 to 100, SIL 2 rating reduces risk by factor 100 to 1000, SIL 3 rating reduces risk by a factor of 1000 to 10,000, and SIL 4 rating reduces risk by a factor of 10,000 to 100,000.

Designing Safety Instrumented Systems (SIS) involves different companies and organizations. The work is split between engineering contractors, producers and vendors and all their activities require to be supervised by one responsible person—the SIS authority. The methodology to determine the SIL requirements includes: identification of barrier elements to reduce risks related to use and operation of the associated equipment, defining the required instrument based protective systems, determining the Safety Integrity Level (SIL) for Safety Integrated Functions (SIF) performed by Safety Integrated Systems (SIS), and description of the minimum SIL/probability of failure on Demand (PFD) requirements.

SIL determination methods depends on the complexity of the process scenario under consideration, the severity of the consequences, the required risk reduction and available information on parameters relevant to process risk. SIL can be determined using the following methods: Risk Graph/Risk Matrix, Quantitative Assessment, Layers of Protection Analysis (LOPA).

In Oil and Gas industry safety integrity level is often determined according to formulas included in IEC Standards 61508 and 61511 following Norwegian Oil and Gas Association Guideline No 070 Application of IEC 61508 and IEC 61511 in the Norwegian Petroleum Industry. Failure rate of system components is calculated based on formulas from IEC Standard 61508, e.g.

$$MTBF = MTTF + MTTR$$

where

MTTBF—mean time between failures
MTTF—mean time to failures

MTTR—mean time to repair/restoration

$$\lambda_{\text{tot}} = \frac{1}{\text{MTTF}}$$

λ_{tot}—total failure rate

$\lambda_D = \lambda_{du} + \lambda_{dd}$

λ_D—dangerous detected and undetected failure rate

λ_{du}—dangerous undetected failure rate

λ_{dd}—dangerous detected failure rate

$\lambda_S = \lambda_{su} + \lambda_{sd}$

λ_S—safe failure rate

λ_{su}—undetected safe failure rate

λ_{sd}—detected safe failure rate

$\lambda_{\text{tot}} = \lambda_{su} + \lambda_D + \lambda_{\text{non effect}}$

$\lambda_{dd} = DC * \lambda_D = DC_D * \lambda_D$

$$DC = \frac{\lambda dd}{\lambda D} = \frac{\lambda dd}{\lambda du + \lambda dd}$$

DC—diagnostic coverage

$$\lambda_{du} = (1 - DC) * \lambda_D$$
$$\lambda_{sd} = DC_S * \lambda_S$$
$$\lambda_{su} = (1 - DC_S) * \lambda_S$$

Modern maritime control and monitoring systems are complicated and consist of many control loops. In addition, mathematical formulas used for SIL determination are complicated which is why consulting companies and control and monitoring certifying bodies are using dedicated software for SIL determination.

7.2.4 Kongsberg K-Safe SIL1, 2, 3 solutions

The Kongsberg computerized safety system K-Safe[3] collects all safety related information into a common system for automatic or manual processing. These safety systems are additional to control systems in order to increase safety onboard by providing a better operator overview of the onboard safety situation. K-Safe may include Operator Stations (OS), Remote Controller Units (RCU), Remote Input/Output Cards and the communication system. Some of the K-Safe elements are certified by TÜV.

Kongsberg K-Safe 3 complies with IEC 61608 SIL 3 and has 1oo2 redundant system architecture, redundant controllers/computers, I/O (RIO), network and power. This system has full diagnostics on all safety critical functions and is used for Emergency Shut Down (ESD), High Integrity Process Shut Down

---

3. Kongsberg webpage https://www.km.kongsberg.com/ks/web/nokbg0397.nsf/AllWeb/007C8A66 DC2782F7C1257B9600247725/$file/350840a.pdf?OpenElement.

(PSD) applications. In that system both controllers are active and may shut down any process independently. Failure in one controller does not prevent shutdown function by other controller. The K-Safe 3 system allows hot replacement of all redundant components without shutting down the controlled processes.

Kongsberg K-Safe 2 complies with IEC 61608 SIL 2 and has 1oo2 redundant system architecture, redundant controllers and single I/O (RIO). In that system both controllers are active and may shut down any process independently. Failure in one controller does not prevent shutdown function by another controller. Failure in one output card shuts down all outputs of this card. The K-Safe 2 system allows hot replacement of all redundant components without shutting down the controlled processes.

Kongsberg K-Safe 1 complies with IEC 61608 SIL 2 and has 1oo1 system architecture, single controller and single I/O (RIO). In that system both controllers are active and may shut down any process independently. Failure in the controller or I/O (RIO) causes system shutdown. K-Safe 1 system allows hot replacement only for the input cards.

Chapter 8

Application of maritime legislation, guidance and standards on automation systems

Chapter outline

Abstract

Having in mind the large quantity of potentially applicable maritime legislation e.g. Conventions, Codes, Guidance and Standards it is necessary to provide guidance regarding their application on vessel and offshore unit control and monitoring systems. This describes the necessarily of applying SOLAS Convention, MODU Code, ISO Standards, IEC Standards, European Union Directives, Coastal State rules for Offshore Units, Norwegian Maritime Authority Requirements, UK Health and Safety Executive Regulations, US Coast Guard and API Recommended Practices and Standards, and NORSOK Standards.

8.1 General

Previous chapters outline the large number of regulations, requirements and standards that ship and offshore unit automation systems may possibly be required to comply with.

Appendix 1A and Appendix 1B contains a typical list of contractual regulations, requirements and standards usually included in Technical Specifications, Vendor Quotations or Technical Agreements, that are obligatory for ordering, installation, and commissioning of automation systems for ship and offshore units.

Ship and Mobile Offshore Unit Automation. https://doi.org/10.1016/B978-0-12-818723-4.00008-9

8.2 Maritime legislation application

The main regulations to ships operating in international waters are contained in an international maritime treaty—the IMO Safety of Life at Sea (SOLAS) Convention which sets minimum safety standards for the construction, equipment and operation of ships. The convention is obligatory to all signatory flag states. The convention is enforced by the Contracting Governments agencies and authorized Recognized Organizations (RO). Such organizations are often Classification Societies that carry out ship surveys and issue Ships' Safety Certificates on behalf of the Administration—the Government of the state whose flag the ship is entitled to fly. The SOLAS Convention regulations do not apply to ships of war and troopships, cargo ships of less than 500 gross tonnage, ships not propelled by mechanical means, wooden ships of primitive build, pleasure yachts not engaged in trade and fishing vessels. The Administration may permit other solutions than required by the SOLAS Convention, provided they are at least as effective as that required by the regulations.

Main regulations for offshore units are contained in the IMO Code for the Construction and Equipment of Mobile Offshore Drilling Units (MODU Code). An Offshore Unit for which a Maritime Administration issues a Mobile Offshore Unit Safety Certificate must comply with MODU Code requirements. The Code allows documented exemptions that are adequate to the Code requirements and documented particular design equivalents at least as effective as that provided in the Code.

8.3 Classification societies rules application

The financial sums involved in loss of a ship or offshore unit loss, third party loss or ship cargo loss are very high, and vessel owners need to underwrite those losses. Owner pay a premium to Underwriters or Protection and Indemnity Insurance Clubs (P&I Clubs) depending on the risk involved with the vessel.

To limit losses of ship/offshore units, the involved Underwriters and P&I Clubs require not only Ship or Unit Safety Certificates confirming fulfillment of the requirements of the SOLAS Convention or the MODU Code, but also a ship/offshore unit Class Certificate issued by a Classification Society confirming fulfillment of Classification Society Rules. The different Classification Societies develop and publish their own rules.

8.4 International standards application

Requirements included in many international standards, for example International Organization for Standardization (ISO) and International Electrotechnical Committee (IEC) are not obligatory unless particular standards are imposed by Maritime Administration directives, Classification Societies Rules or are listed in the owner's Technical Specification for the vessel. In such cases, the referenced ISO, IEC Standards are obligatory for the contracting parties.

8.5   European Union directives

European Union Directives are obligatory for EU flagged ships and offshore installations in EU waters.

8.6   Maritime administration requirements

In addition to IMO SOLAS Convention and MODU Code, the local Maritime Administrations sometimes impose requirements that are obligatory in their waters, for example Coastal State rules for Offshore Units, Norwegian Maritime Authority requirements, UK Health and Safety Executive Regulations, US Coast Guard and American Petroleum Institute (API) Recommended Practices and Standards, and NORSOK Standards. Some of these requirements are imposed by the vessel owners.

Chapter 9

Designing control and monitoring systems

Chapter outline

Abstract

This chapter describe the design phases of control and monitoring systems including Conceptual design, Basic design, Detail engineering, technological documentation and provision of operating instructions and As-built documentation. Additionally the chapter lists different types of documents typically produced in each design phase.

9.1 Conceptual design

Basing on their own competence and experience shipyards, designing/consulting offices or Owner's Technical Departments prepare conceptual designs for new vessels or offshore units or smaller electrical installations that take into account project requirements, technical feasibility with regard to existing capabilities, economic value and market trends. A conceptual design package usually includes:

- technical specification,
- principal diagrams,
- necessary plans,
- P&ID diagrams,
- General Arrangement drawings (GA),
- equipment arrangements,
- equipment Makers List.

Conceptual design defines the main/general parameters of a new vessel or offshore unit, which should not change in the subsequent basic and detailed

Ship and Mobile Offshore Unit Automation. https://doi.org/10.1016/B978-0-12-818723-4.00009-0

design phases. During this stage the vessel control stations are defined and equipment weights are estimated. The conceptual design work is normally completed prior to entering into the main contracts for a new project. This designing phase is usually performed by designing/consulting offices.

9.2   Basic design

The basic design phase follows the conceptual design phase. It is usually carried out after award of contract for new build or conversion, but in some projects an Owner may pay for this scope prior to contract award if the bidding shipyards are not felt to be competent for this task. The basic design tasks are carried out by the shipyard design office, or by subcontracted independent design offices. The basic design work is based on the conceptual or contract design, or other documentation defining the desired vessel/offshore unit or installation.

The scope of basic design work is much more detailed than conceptual design. The basic design phase is developed according to the technical requirements, Classification Rules and other regulations and standards listed in the contract Technical Specification. The minimum requirements of a project are outlined in the Classification Rules, but usually the basic design work must address additional Owner's requirements including some which may be described in philosophies i.e. Vessel/offshore unit automation management, power management, dynamic positioning (DP), electrical operation, emergency shutdown (ESD), fire and gas, cable specification and selection, automatic system overview etc. Additional information may be included in control and monitoring systems functional descriptions, design specifications, configurations, data sheets and systems overview developed for Dynamic Positioning system, Power Management system, Propulsion and Control systems, Alarm and Monitoring system, Human Machine Interface (HMI) display etc.

Basic design work is normally required to be approved by Owner's representative, Classification Society and vessel's Flag State Administration (where applicable). After having approved basic design documents the shipyard design office can proceed to the detailed design phase. At this stage or earlier, the shipyard procurement department will be able to use confirmed design information to prepare reliable Request for Quotation (RFQ) or Information to Tender (ITT) documents to be sent to equipment vendors and subcontractors.

The basic design phase is usually performed by design/consulting offices, sometimes with the cooperation of the vessel or offshore unit owner.

Examples of basic design phase drawings/documents are presented in Appendix 2.

9.3   Detail engineering phase

The detailed engineering phase is developed by the shipyard designing office or shipyard subcontractor and takes into consideration the basic design approved by the following parties: Owner, Classification Society and vessel's Flag State Administration (where applicable). The scope and detail of this engineering phase depends on the shipyard's production technology and methods, and the information needs of the workers or subcontractors employed. For electrical control and monitoring systems, detailed engineering deliverables include working documents such as P&ID drawings, Input/Output (I/O) alarm and monitoring list, schematic and connection diagrams, cable schedules, cable plans, cable tray routing, equipment arrangement drawings, work instructions and other production drawings.

The detailed engineering design phase is usually performed by the ship-builder's design office and his subcontractors.

Examples of detail engineering drawings/documents are presented in Appendix 2.

9.3.1 Drawing numbering

Project drawings are numbered according to the shipyard's internal system, sometimes with an additional set of numbers if required by the Owner's system. The most widely used numbering system used by shipping and offshore companies, shipyards and consultants worldwide is the SFI Group System developed by Ship Research Institute of Norway and now owned by SpecTec.[1] This standard provides a functional subdivision of technical ship or rig information. In this system, electrical, instrumentation and communication systems drawings are numbered as below:

42—Communication Equipment
    421—Radio Plant
    422—Lifeboat Radio Transmitters, Emergency Radio, Direct Finder
    423—Data Transmission Plants (Telex), General Purpose EDB Plants
    424—Very High Frequency (VHF) and Ultra High Frequency (UHF) Telephones, etc.
    425—Calling System, Command Telephone, Telephone Plants, etc.
    426—Speaking Tubes, Tube Post Plants etc.
    427—Light and Signal Equipment—Lanterns, Whistles etc.

79—Automation Systems for Machinery
    791—Manoeuvre Consoles, Main Consoles
    792—Common Automation Equipment (Engine Room Alarm etc.)
    793—Automation Equipment for Propulsion Machinery and Transmissions
    794—Automation Equipment for Boilers etc.
    795—Automation Equipment for Diesel/Turbo Aggregates etc.

1. See web page http://www.spectec.net/technical-coding-solution.

797—Automation Equipment for Other Machinery Components.
798—Cables/Leads and Piping for Automation Systems for Machinery

85—Electrical Systems General Part
86—Electrical Power Supply
   861—Generators/Alternators
   865—Transformers
   866—Batteries and Chargers
   867—Converters and Rectifiers
   868—Shore Supply
87—Electrical Distribution Common Systems
   871—Main Switchboard
   872—Emergency Switchboard
   873—Group Starters
   874—Local Starters
   875—Distribution Panels and Boards
88—Electrical Cable Installation
   881—Cable Trays and Installation in Pontoons and Columns
   882—Cable Trays and Installation in Accommodation, Superstructure etc.
   883—Cable Trays and Installation of Drilling Machinery on Main Deck
89—Electrical Consumers
   891—Electrical Lighting Systems for Machinery in Pontoons and Columns
   892—Electrical Lighting Systems in Accommodation, Superstructure etc.
   893—Electrical Lighting Systems for Drilling Machinery on Main Deck
   897—Electrical Ventilators
   898—Electrical Motors

9.4   Technological documentation—installation and commissioning instructions

Technological documentation is prepared by the design offices to assist the installation and commissioning of control and monitoring systems onboard. Usually shipyards and their subcontractors perform installation and commissioning work according to written procedures and standard practices, for example there may be a shipyard Standard Practice for Electrical Installation. For complicated systems, there are often prepared additional technological documents including system/equipment drawings, mounting instructions, commissioning manuals and equipment catalogue cards.

Technological documentation is usually prepared by the shipyard design/technological office or its subcontractors.

Examples of technological documentation are presented in Appendix 2.

9.5   As-built documentation

When a project is finally built and commissioned, the shipyard design office or their subcontractor prepares As-Built documentation includes updated and corrected drawings and documents from basic design, detailed engineering, together with copies of the equipment manuals and operating instructions for the vendor supplied items installed onboard.

As built documentation is usually prepared by shipbuilder designing office or its subcontractors.

Chapter 10

Procuring control and monitoring systems

Chapter outline

Abstract

In this chapter is described the process of specifying and purchasing control and monitoring systems and equipment. Additionally the chapter includes examples of Integrated Automation and Control System (IACS) Configuration drawing, Technical Specification, the scope and form of typical System Functional Descriptions, form of typical Piping and Instrumentation Diagrams (P&ID), and examples of the symbols often used on P&ID Diagrams, together with a few examples of Mimic Diagrams.

Ship and Mobile Offshore Unit Automation. https://doi.org/10.1016/B978-0-12-818723-4.00010-7

10.1   General

During the procurement phase the vessel Owner or the engineering company authorized by him orders equipment or subcontracts for the new vessel. In order to perform this task, documents must be prepared describing in detail the scope of Contract/Services to be supplied. Such documents are completed with attachments containing the necessary technical data defining the ordered equipment or services, i.e.:

– Vessel/Offshore Unit Technical Specification,
– Systems Functional Descriptions or System Philosophies,
– Piping and Instrumentation Diagrams (P&ID),
– Necessary plans, equipment arrangements, manuals etc.

Such document packages are called Information to Tender (ITT) or Request for Quotation (RFQ). Having received all necessary information, the bidding vendor or contractor prepares a 'Tailor made' offer or proposal that includes technical and commercial parts, delivery schedule and Quality Assurance part.

10.2   Vendor quotations

Basing on the received Information to Tender (ITT) or Request for Quotation (RFQ) the subcontractors or vendors prepare 'Tailor made' quotation and develop and detailed scope of the offered equipment and services. The quotation shall address all requirements stated in the Information to Tender (ITT) or Request for Quotation (RFQ), unless any specific exclusions are mentioned.

A typical quotation includes the following information: technical data, scope of supply, exclusions, budget price, warranty conditions, options with prices, training costs, sales conditions, delivery time, offer's validity time, insurance conditions, limitation of liability, confidentiality requirements, place of jurisdiction and applicable law. Sometimes the bidding vendor or subcontractor also provides a Compliance Matrix listing the requirements not covered by the quotation or covered only partially, and giving explanations why he cannot fulfill the specified requirements. Before signing the Contract, the Compliance Matrix is agreed between parties, and this will form part of the overall agreement.

A quotation for Integrated Automation and Control System (IACS) typically includes a System Configuration drawing. An example of an IACS Configuration drawing is shown in Fig. 10.2.1.

In addition to the System Configuration drawing, a quotation for Integrated Automation and Control System (IACS) will normally contain[1] descriptions of the functions and hardware to be supplied.

1. Based on Kongsberg IACS Offer

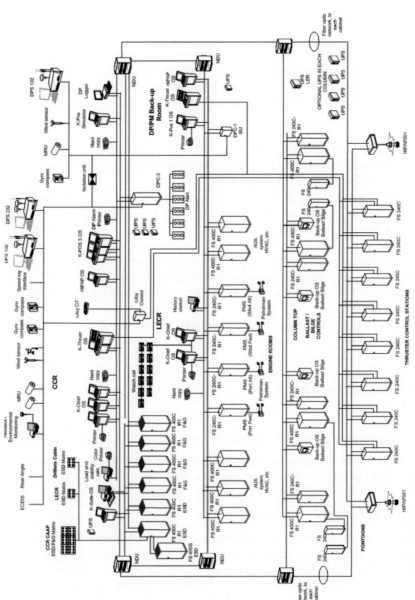

FIG. 10.2.1 Integrated Automation and Control System (IACS) Configuration drawing. *(Based on Kongsberg IACS Offer.)*

Integrated Automation and Control System (IACS) functions typically incorporate:

- Alarm system including Watch Call system and Engineer Fitness system, Hospital Call and Refrigerator Alarm system,
- Monitoring system,
- Control system including Power Management System (PMS) and machinery management with remote control of machinery pumps, PID controllers, remote control of valves, tanks level gauging, automatic bilge control, sequential start/stop of propulsion auxiliaries.

Hardware descriptions typically cover:

- Operator Stations (OS) with marine computers, colour display(s), operator panels with trackball, dual process network connections, administrative network connections, power supply,
- Field stations with redundant real-time processors working in Master/Hot standby system, dual net interface, dual I/O interface, I/O bus galvanic isolation, network switches,
- System interfaces quantity for hardwired input/output signals and serial input/output signals, I/O modules with analogue inputs 0-20mA/0-10V, analogue outputs with analogue inputs 0-20mA/0-10V, analogue inputs PT100, analogue inputs potentiometer, digital inputs, digital outputs and pulse/frequency inputs,
- Processes network connecting all processes network distribution units/controllers/hubs via fibre optic cables for connections between remote locations and Local Area Network (LAN) shielded twisted pair CAT7 cables used locally,
- Watch call system panels,
- Engineer fitness panels/pushbuttons,
- Printers,
- Uninterruptable Power Supplies (UPS).

Additionally, the quotation typically includes the number of Graphical User Interface to be installed in the Operator Stations (OS).

10.3 Technical agreement (TA)

After receipt and negotiation of the Quotation, an agreed Technical Agreement is signed by shipyard/owner representatives and subcontractor in which commercial and technical details, including compliance matrix, are agreed between parties.

The Technical Agreement usually includes the following sections:

- General—containing buyer and seller details, applicable law and regulations, delivery time with delay penalties, force majeure causes definitions, guarantee conditions,

- Basic technical specification—containing vessel/offshore unit class, flag, list of applicable rules, statutory regulations and standards, applicable certificates, vessel/offshore unit necessary spare parts, documentation scope and approval procedure, testing requirements,
- Extracts from overall vessel/offshore unit Technical Specification applicable to the specific Technical Agreement scope,
- Scope of delivery in form of agreed earlier Quotation,
- Quality Assurance plan,
- System name(s) and type(s),
- Seller and Buyer names and signatures.

10.4 Vessel/offshore unit contract and technical specification

To build/convert the vessel or offshore unit or some parts of her onboard systems or installations, the ordering company (usually the vessel Owner) signs a legal agreement or contract with the selected contractor who is then responsible to fulfill the contract obligations. The contract package usually consists of an Agreement, Conditions of Contract and Appendices/Annexes or Exhibits.

Conditions of Contract usually include the following sections: General Provisions, Performance of the Work, Progress of the Work, Variations, Cancellation and Suspension, Delivery and Payment, Breach of Contract, Force Majeure, Liability and Insurances, Limitation and Exclusion of Liability and other provisions.

The Technical Specification is usually included as an Appendix or Exhibit to the Conditions of the Contract. The Technical Specification is usually developed by Owner's Technical Department or by its Consulting office, and is negotiated and agreed together with the yard or subcontractor. Such specifications are structured in different chapters sometimes following the previously mentioned Norwegian SFI numbering system, although other systems are also used. Regardless of the exact structure, an overall vessel specification will include the following subjects:

- Part I—General Provisions—containing a description of the Vessel/Offshore Unit—main dimensions, deadweight and capacities, power and speed, fuel consumption, freeboard, technical guarantee, applicable rules and regulations—classification, flag, certificates, properties of the vessel, inspection, testing and trials, documentation and dry docking.
- Part II—Hull Construction and Corrosion Protection,
- Part III—Accommodation—including requirements for furniture and decoration, sanitary outfit, stores, windows, doors, stairways, air conditioning and ventilation.

- Part IV—Outfitting—covering anchoring and mooring appliances, steering system, cargo hatches, lifesaving appliances, navigation equipment, deck outfitting, firefighting systems and hull piping systems.
- Part V—Machinery—covering main engines, shafting and propellers, electric generating plant, compressed air systems, cooling water systems, feed and drain systems, fuel oil system, lubricating oil system, fire, bilge, ballast and general service systems, fresh water generator, fresh water service system, fans, auxiliary machinery, piping systems in engine room, automation, monitoring and remote control including control stations, control room, control of electric generating plant, control of purifiers, compressors, remote quick closing valves and ballast control.
- Part VI—Electrical—wiring, primary power source-main generators, main switchboard, emergency generator, emergency switchboard, shore connection, secondary source-transformers, storage batteries, electric motors and starters, lighting system, navigation and signal lights, interior communication, alarm and measuring system, battery phones, automatic telephone system, hospital call system, bridge watch monitoring system, engineer call system, nautical equipment, radio communication equipment.

There may be a separate chapter devoted to special Mission Equipment (e.g. drilling equipment) or the cargo handling system.

Another example of Technical Specification structure is as follows:

- Chapter 1—General—main particulars of the vessel, performance, trial speed, class, tonnage, regulations, certificates, Owner supplied equipment.
- Chapter 2—Hull—hull material, engine room area construction, deckhouse and superstructure, hull outfitting and material external, internal protection.
- Chapter 3—Cargo Equipment—hatches, deck cranes, loading/discharging liquid systems.
- Chapter 4—Maneuvering Machinery and Equipment, bow thrusters, stabilising systems, Dynamic Positioning, navigation equipment, communication equipment, anchoring, mooring and towing equipment, engine room lifts.
- Chapter 5—Equipment for Crew and Passengers—lifesaving, protection and medical equipment, insulation, panels, bulkheads, doors and windows, external deck covering, ladders, stairs, gangway, furniture, inventory, entertainment equipment, galley/pantry, provision plants, laundry equipment, ventilation, air-conditioning and heating systems, sanitary systems, black and grey water.
- Chapter 6—Machinery Components—diesel engines for propulsion, diesel generators, propellers, boilers, steam generators, emergency generator.
- Chapter 7—Machinery Systems—fuel system, lube oil system, cooling systems, compressed air systems, exhaust gas and air intakes systems, automation systems for machinery—engine room console, bridge consoles, dynamic positioning consoles, integrated automation system, alarm and monitoring system.

– Chapter 8—Ship Common Systems—ballast and bilge systems, fire detection, fire-fighting systems, electric and electronic systems—electric distribution systems, main switchboard, emergency switchboard, power management system, computer LAN network, transformers, batteries and chargers, shore supply, starters, frequency converters, distribution boxes, electric cable installation, electric lighting systems, electric motors.

Other technical Appendixes or Exhibits to Conditions of Contract are usually the General Arrangement Plan and the agreed Makers List.

In some cases, the Technical Specification and General Arrangement Plan are developed by the shipbuilder design office and are negotiated and agreed with the Owner's Technical Department. Such a situation is typical when a shipbuilder is building standard ships in long series and has developed a standard design and specification for discussion with potential owners.

10.4.1 Scope and form of typical technical specification— Ship's processes monitoring, control, regulation and automation

An example of typical specification language is given below:

---

79 Monitoring, Control, Regulation and Automation[2]

1. General provisions

Monitoring, Control, Regulation and Automation Systems shall comply with Classification Society requirements for unmanned engine room. The system shall be based on computerised system.

Necessary redundancy shall be achieved in various systems according to class requirements.

Vessel automation system to be divided into two subsystems: one for Engine Control Room & Alarm and Monitoring system and second one for Ballast/deballast Control and Monitoring system.

All electronic equipment shall be protected against sustained and transient over voltage and electromagnetic interferences.

Standard and well proven equipment shall be used.

Total number of channels for automation and monitoring systems is about 9000.

The Mimic diagram shall be provided for propulsion system, PMS, systems pumps control, diesel engines, valves control. Total number of Mimic's is about 50 pcs.

Monitoring, Control, Regulation and Automation Systems shall be supplied from an independent 24V UPS system with battery duration over 30 minutes.

Alarm systems, control systems and shut down systems shall be separate as far as possible, with minimum of common components and with separate power supplies. Interconnections between different systems shall have galvanic isolation.

Continued

---

2. Based on Kongsberg project

Redundant machinery shall have independent control systems, including separate power supplies.

Automatic control systems shall have provisions for manual control and shall have MAN/AUTO switches.

All automation cabinets shall have the same keys as far as practicable.

2. Integrated Automation System

Integrated Automation System (IAS) consists of Alarm and Monitoring system, Remote Control systems and shall be independent of propulsion remote control systems.

Computer based Integrated Automation System shall be installed according to rules requirements for unattended machinery room.

IAS shall include the following functions: Alarm and monitoring all engine room machinery, and Remote control of all engine room machinery including pumps and valves, start/stop, stand-by, start after black out functions and speed control where applicable.

The main components of IAS are:

— *Workstations*

Workstation in Engine Control Room and on bridge shall include: two LCD display, two keyboards, trackball and PC, one color printer for alarms and one printer for data logger.

— *Extension Alarm System & Dead Man Alarm*

An extension alarm system shall be provided to alerts the officers when an alarm is detected by alarm and monitoring system. Change over switch shall be installed in engine room to set up ER attended/unattended and this need to be accepted on the bridge. When an alarm is activated and not acknowledged in three minutes an alarm signal shall be initiated in engineer's cabins. In attended engine room mode an alarm shall be initiated in engine control room and in engine room.

Extension alarm panels shall be installed on the bridge, engine control room, Chief Engineer room, 2nd, 3rd, 4th Engineer room, Electric Engineer room, Officers Mess room, Crew Mess room, recreation room, conference room.

Extension alarms on the bridge shall be arranged in five groups: Group 1—Fire alarm, Group 2—Propulsion Plant, Electrical Plant, Group 3—Ship no-essential service, Group 5—IAS system failure.

Dead Man Alarm shall be initiated automatically and six (6) pushbuttons to reset an alarm shall be installed in engine room and one in control room.

— *Engine Room Alarm Indication System*

Engine Room Alarm Indication System shall be installed and shall include: Machinery alarm, CO_2 alarm, General alarm, Fire alarm, Telephone alarm, Dead Man alarm and calling signal.

Light panels shall have colored light fields corresponding to each alarm. Panels shall be provided with sounder and flashing light.

— *Piping systems*

Pumps, machinery and valves shall be remote controlled as far as practicable through IAS by VDU's and keyboard in Engine control room console.

Valves shall be operated by electro-hydraulic actuators with possibility for local manual control. Valves position feedback shall be signaled using mechanical limit switches.

At least the following systems shall be provided with remote control facility with status indicating in IAS: fuel oil filling, transfer system, lubricating oil system, fresh water system, sea water system, ballast water system, bilge water system, firefighting system.

For above systems Mimics shall be provided on LCD display.

– Pumps control

Installed in Engine room duplicated pumps shall automatically change-over in abnormal situation.

The pumps shall be started either manually or automatically from Workstation MIMIC. Workstation shall have possibility to change-over between main or stand-by equipment control mode.

Stand by functions shall be activated in automatic control mode.

Pump motors shall start automatically on loss of discharging pressure or voltage of the operating one, and stand-by starting alarm shall be initiated on IAS alarm system.

Pressure switches shall be installed on each outlet of double pumps.

All pumps controlled by IAS shall be sequentially restarted from IAS after black-out.

On VDU the following pump status shall be indicated: Running, Stand-by when pump is ready to use, Auto/Manual mode, Local/Remote control, Failure alarm.

From VDU shall be possible pumps remote Start/Stop.

– Temperature, Pressure and Viscosity control

Control valves and other control elements shall be equipped with electrical or pneumatic actuators. In case of failure, the valve shall go to fail safe position or not changed position.

Pneumatic actuators and 3-way valves shall be equipped with means for manual control to by-pass remote control.

– Oil separators control and automation

The Fuel oil and Lubrication oil separators and their heaters shall be fully automatic with their own sludge discharge. Cleaning program shall start only when separator is stopped. Failure of separators shall be signaled in IAS Alarm system.

– Steam plant control

Steam plant shall be fully automatic with pressure control and waste heat shall be used with maximum efficiency.

Oil fired boiler shall start automatically when the steam production from waste heat—Exhaust gas boilers is below the limit and shall stop automatically when steam is not needed.

Steam heated services shall have automatic temperature control, locally mounted and direct acting.

Salinity detectors with alarm shall be provided.

– Fired boiler control

Fired boiler automatic control shall be independent and protection and alarm displays shall be arranged in free standing panels clearly visible from boiler front.

The combustion control equipment shall be arranged to control the oil flow to the burner and the oil-air ratio.

Continued

Fired boiler group alarm, running signal and level indication shall be connected to IAS.

Fuel oil shall be shut-down in case of: low/very low water level, ignition failure, flame failure, low fuel oil temperature/pressure, fuel oil pump stop, electric failure, forced draft fan stop.

   − *exhaust gas boilers control*

Exhaust gas boilers steam temperature shall be monitored by IAS.

Generated by exhaust boiler steam shall be led to the atmosphere.

   − *Miscellaneous systems control*

All auxiliary systems and units shall be controlled automatically also when are not described in this Technical Specification e.g. air compressors, refrigerating plant etc.

   − *Fresh water generator control*

The fresh water generator shall be arranged for fully automatic operation.

Salinity alarm systems shall be included with dump of water in case of high salinity.

   − *Sanitary water control*

Fresh water hydrophore pumps shall be controlled by pressure switches installed on hydrophore receiver.

Low pressure alarm shall be signaled in Alarm system.

Flow-meter with local reading shall be installed for measuring water from a shore.

   − *Waste oil incinerator*

Waste oil incinerator control shall be independent and shall have automatic burner/combustion system and safety incinerator system.

Group alarm shall be signaled in Alarm system.

   − *Sewage system*

Discharge pumps shall be stopped by low level switch and separate level switch shall be installed for high-high sewage level.

Group alarm shall be signaled in Alarm system.

   − *Bilge water separator*

The bilge water separator shall be equipped with automatic control and monitoring and recording quality of the water after separator. If the oil content is higher than permitted than three-way valve leads water after separator back to bilge water tank. Position of three-way valve shall be indicated locally.

Group alarm, including high oil content after separator and low level in bilge water tank shall be signaled in Alarm system.

   − *Control of safety related tanks and piping systems*

Remote control of safety related tanks and piping systems shall include pumps start/stop, valves control, tank levels indication etc.

Between watertight compartments shall be installed remotely controlled isolation valves.

Remote control shall include the following systems: ballast system, bilge system—including bilges normally and emergency discharge and separate high level alarm for each bilge, fire systems—including firefighting, deck wash and general services, fuel oil—including bunkering, pumping fuel ashore and between fuel tanks, high level alarm in fuel oil storage, setting, service tanks, protection against overbunkering, fresh water and sanitary systems—including bunkering, pumping fresh water between water tanks, high level alarm in fresh water tanks, protection against overbunkering, sewage system—including sewage discharge overboard or ashore to the barge, sludge system—including sludge discharge to the barge.

— Level measurement

Level measurement in meters and tank capacity in cubic meters shall be fitted in the following tanks: heavy fuel oil storage tanks, oil settling tank, service tank, and overflow tank, marine diesel oil storage tank, oil setting tank and service tank, all water tanks, fresh water tank, distilling water tank, potable water tank, bilge water tank.

In all ballast tanks two sets of remote sounding shall be installed.

— Valve control

The valves control system shall be equipped with electro-hydraulic actuators.

The ballast system shall be remotely controlled from the MIMIC's on workstation LCD. On MIMIC's valves status closed/open shall be indicated. The indication shall be initiated by proximity switches with IP68 enclosure.

De-ballast compressors shall be included in sequence start after black-out.

— Bunkering

Bunkering system operations shall be controlled from ECR. The following equipment shall be provided in Bunkering station: Sound powered telephone and automatic telephone, Manifold pressure gauges, Fire alarm push-button, Discharge pump stop push-button.

3. Control consoles

In Bridge, main console the following equipment shall be mounted: propulsion control panel, control lever and instruments, emergency telegraph, maneuvering recorder, IAS work station, alarm system panel, sound powered telephone and automatic telephone etc.

In Bridge front wall panel, the following equipment shall be mounted: rudder angle indicators, shaft RPM indicators, propulsion motor indicators etc.

On the bridge wings, the following equipment shall be mounted: rudder angle indicators, shaft RPM indicators. In bridge wing consoles the following equipment shall be mounted: propulsion control panel, control lever and instruments, shaft RPM indicators, propulsion power indicators, power control panels.

In Engine Control Room Console the following equipment shall be mounted: Two IAS workstations with LCD display, keyboard and trackball, One color printer for alarms and One printer for data logger, Two propulsion MW meters, Two propulsion RPM meters, Two propulsion emergency stop push-buttons, Two steering gear indicators, One Main Engine (ME) Emergency stop push-button, Four Auxiliary Engines (AE) Emergency Stop push-buttons, Four Main generator kW meters, Twenty four analogue measuring instruments for essential pressures and temperatures, Clock, Main engine oil viscosity indicator, Sound powered telephone and automatic telephone, Fire alarm repeater, Speed log indicator, Steering gear alarm panel, Emergency stop push-buttons, Indication lamps etc.

Continued

Engine Control Room Console (ECRC) shall be supplied from: MSB—3x220V, 60 Hz, ESB—3x220V, 60 Hz and from Batteries 24V, DC.

Main supply failure shall cause automatic transfer to emergency supply and shall initiate alarm.

4. Local instrumentation

All machinery, including piping systems, shall be equipped with local instrumentation as follows:

— *Pressure gauges on the suction and pumps discharge, on the inlet of coolers, on pressure valves, after pressure reduction valves etc.*
— *Pressure gauges at the cooling water and lubricating inlets,*
— *Thermometers in the inlet/outlet for coolers, heaters in the oil and cooling water outlets.*

5. Sensors

On the vessel shall be used standardized sensors.

Analogue pressure sensors with transmitters shall be used as far as practicable.

Sensors for high temperatures for example for exhaust gas, shall be NiCr/ Ni thermocouples and for low/intermediate temperatures shall be thermistor or PT100 types.

Analogue pressure sensors shall be used for ballast tanks. Float type level switches can be used where switching is required.

Level switches for highly corrosive contents and sewage shall be of stainless steel.

All temperature sensors to be installed in pockets of protecting material.

Shut off and test valves with standard connection shall be installed for all presostats and pressure measuring sensors.

Proximity switches are preferred, instead switches with moving parts.

6. Automation systems electric power supply

Automation equipment shall have redundant power supply with automatic change over and alarm. Duration of supply shall be according to rules requirements but not less than 30 minutes.

10.5 Systems functional descriptions

In the early stages of project contract negotiations not all technical requirements are finalized, so the Technical Specification is not always detailed enough to provide all information required to design control and monitoring systems. The necessary information can be detailed in separate Systems Functional Descriptions or System Philosophies. Such documents specify the functions that the system or equipment must perform. This type of document usually describes what is needed by the system user, and the required properties of inputs and outputs. Sometimes the System Functional Description is called a System Philosophy.

10.5.1 Scope and form of typical system functional description—sea water cooling system

Below is an example of typical text found in a System Functional Description[3];

---

1 Introduction

1.1 General

The control and monitoring of Sea Water Cooling System shall be included in the Integrated Automation System (IAS).

The Sea Water Cooling System consists of thruster cooling system, central fresh water cooling system and the seawater service system.

1.2 Classification

The Unit complies with the DNVGL Offshore Standards with the following class designation;

✠1A1 Column Stabilized Drilling Unit (N), HELDK, E0, DYNPOS AUTRO, POSMOOR ATA.

1.3 Rules and Regulations

- *DNV-OS-202 Automation, Safety, and, Telecommunication Systems, Part 6 Chapter 3 Periodically Unattended Machinery Space*
- *DNV OS-D101 Marine and Machinery Systems and Equipment*
- *IMO 2009 MODU Code*
- *NMA regulations*

2 References

- *P&ID, Sea Water Cooling System*
- *P&ID, Sea Water Service System*
- *P&ID, Fresh Water Cooling System*
- *Functional Description, Fresh Cooling Water System*
- *P&ID, Hydraulic Valve Oil System*
- *Functional Description Power-packs Valve Cabinet*

3 Design data

Inlet sea water temperatures vary between 0° and 32°C.

The system is designed to provide redundancy with respect to single failures and with consideration to physical hazards as fire and flooding.

3.1. Data of sea water systems

3.1.1 Aft sea water systems, each side
- *Central Coolers 14500 kW (each)*

3.1.2 Forward sea water systems, each side
- *Central Coolers 14500 kW (each)*

3.1.3 Thruster cooling
- *Thruster fresh water cooler 500 kW*

3.1.4 Service pump system
- *Sewage unit 0.07 m^3/h*
- *Needed capacity service pump 450 m^3/h*

Continued

---

3. Based on Kongsberg project

- Continuous consumers 407.87 m³/h
- Max intermittent (Short time) 446.37 m³/h

3.1.5   Fresh water generator feed pump system
- Fresh water generator, 2 off 16 m³/h
- Fresh water generator feed pump,2 off 20 m³/h, 8bar

3.1.6   Burner Boom/Deluge Pump
- Burner Boom/Deluge pump, 1 off 450 m³/h, 16.5 bar

3.2   Data of sea water pumps
Aft sea water system—two sea water pumps in both sides
- Pump head 2.0 bar
- Pump capacity 900 m3/h
  Forward sea water cooling system—two sea water pumps in both sides
- Pump head 2.0 bar
- Pump capacity 900 m3/h
  Thruster cooling systems—one sea water pump for each thruster
- Pump head 2.0 bar
- Pump capacity 80 m3/h
  Service pump system—two service pumps in both side
- Pump head 5.5 bar
- Pump capacity 600 m3/h
  Two fresh water generator feed pumps
- Pump head 8.0 bar
- Pump capacity 16 m3/h
  One Burner Boom/Deluge pump
- Pump head 16.0 bar
- Pump capacity 450 m³/h

3.3   Flow velocities
Max. 4.5 m/s.

4   System description
4.1   Main Sea Water Cooling System
Eight main sea water cooling pumps are installed, two in each pump room for cooling of the fresh water cooling system via the central water coolers. Each pump has 100% capacity. Normal operation is when one pump running and other one is in standby mode in each pump room.

The main sea water pump located in the aft sea water cooling circuits supplies cooling water for the aft central coolers.

The Burner Deluge pump takes suction from sea chest and the sea water are distributed to burner boom cooling system which also include hull side deluge. In case of failure of Burner Deluge pump, the fire pump system shall act as a backup.

The main sea water pumps supplies sea water for the forward central coolers in the area. The sea water outlets from these central coolers are discharged directly overboard.

4.2   Thruster Sea Water Cooling System
Individual cooling system with a thruster sea water pump, a thruster cooler and a thruster fresh water is arranged for each of the eight thrusters and accessories.

The thruster sea water pump supplies the sea water for the thruster cooler in the area, and the sea water from thruster cooler is discharged directly overboard. To avoid under cooling of thruster fresh water cooling system, a three-way valve is arranged in the thruster fresh water cooling circuit.

4.3 Sea water service system
The service system consists of two pumps, one PS and one STB and supplies sea water for machinery and drilling consumers: Steam drain cooler, Sewage unit, FW generation unit.

4.4 Fresh water generator feed pump system
The two fresh water generator feed pumps, each 100%, take suction from the sea chest.
The design allows for maximum distance from sewage outlet.

4.5 Burner Boom/Deluge pump
Burner Boom/Deluge pump supplies water to the burner boom cooling and hull side deluge.
The Burner Boom/Deluge pump takes suction from the sea chest.
In case of failure of Burner Boom/Deluge pump, the fire pump system will act as a backup.

5 IAS Control and Monitoring
5.1 Valves
All major valves in the Sea Water Cooling System are remotely controlled and monitored via IAS. See IO-List for details.
Sea water sea chest valves and overboard valve (if applicable) are provided with a soft emergency closing button from IAS. The pumps are also stopped with closing dedicated discharge valve. There will be a soft button per each cooling circuit.

5.2 Pumps
5.2.1 Main Sea Water Cooling Pumps
There are (8) eight main sea water cooling pumps. The Main Sea Water Cooling pump set is operated in Duty/Stand-by mode. The Duty pump is normally running (in Auto mode). The Switch over is activated in case of electrical fault or Low discharge pressure on the Duty pump.
Normal operation is one pump running in each corner.
The IAS indicates the operating mode of the pump and alarm at start of the back-up pump.
Sea chest valve should be open on the duty pump. Valves are "fail last type".
Pressure transmitters on discharge of each pump are displayed on IAS with High and Low alarm.
Each set of Main Sea Water pumps supply two Central Fresh Water Coolers. The Central Fresh Water cooler has a differential pressure (between inlet and outlet pipeline) that is indicated on IAS with High alarm.
Temperature indications on the discharge of the Central coolers are presented on IAS with High and High/High alarm.
Automatic sequence:
The starting of the pump will be interlock with the opening feedback of the Sea chest valve. The Sea chest valve will be opened prior starting the pump. At running

Continued

of the pump, the discharge valve will open. It will be automatic close of the discharge valve when the pump stopped.

5.2.2 Thruster Sea Water Cooling Pumps

The thruster sea water cooling pumps, one for each thruster fresh water cooler, are remotely controlled via the IAS. The valves can be remotely manually operated from IAS.

Thruster Sea Water Cooling valves have opened, closed feedback, over torque alarm and local/remote signals sent to IAS. IAS sends Open/Close command when in remote.

Differential Pressure transmitters are also provided at the SW cooling inlet and outlet of the Thrusters fresh water cooler. Those Differential Pressure transmitters are shown on IAS with High Alarm.

Automatic sequence:

The starting of the pump will be interlocked with the opening feedback of the sea chest valve. At running of the pump, the discharge valve will open. There will be automatic close of the discharge valve when the pump stopped.

5.2.3 Sea Water Service Pump

There are two seawater service pumps in port and starboard aft pump room for miscellaneous machinery and drilling utilities.

The sea water service pumps supply two SW cooling lines to a common header. One supply line contains a pressure control valve. First pump is set to regulate the SW at 3.0 bar. A pressure transmitter connected to IAS is installed on the main header. A maintenance bypass line is available to allow maintenance of this PCV.

The Sea Water Service pump set is operated in Duty/Stand-by. The Duty pump is normally running (in Auto mode). The switch over is activated in case of electrical fault or low discharge pressure on the duty pump.

There are also pressure transmitters at the discharge of the SW pumps that are connected to IAS with High and Low alarm.

Automatic sequence:

The starting of the pump will be interlocked with the opening feedback of the Sea chest valve. At running of the pump, the discharge valve will open. There will be automatic close of the discharge valve when the pump stopped.

5.2.4 Fresh water generator feed pump system

The two fresh water generator feed pumps, each 100%, take suction from the aft port and aft the fresh water generator feed pump set is operated in Duty/Stand-by. The Duty pump is normally running (in Auto mode). The Switch over is activated in case of electrical fault or discharge Low discharge pressure on the Duty pump.

Pressure transmitters at the discharge of the pumps that are connected to IAS with High and Low alarm

Automatic sequence:

The starting of the pump will be interlock with the opening feedback of the suction valve.

5.3 MGPS System

The sea water systems are protected from Marine Growth by a Marine Growth Protection system.

IAS received failure alarm, and send maximum current request signal when all related pumps are running for aft sea chests.

If more than one pump is running for each sea chest, IAS send maximum current request signal to MGPS control panel.

6 Operation

6.1 Start-up of main sea water system

Below example is starting system port aft in AUTO mode. The sequence is similar for the other corners.

- *Select which pump to be in duty (Both pump needs to be available),*
- *Open sea side valves to both pumps,*
- *Confirm manual inlet sea valves are opened,*
- *Confirm manual overboard valve is opened,*
- *Confirm that inlet and outlet FW cooler manual valve are opened,*
- *If discharge through the HP Mud pump, confirm manual valve is opened,*
- *Start duty pump by selecting AUTO mode,*
- *Discharge valves for the pump that is selected as duty pump will be automatically open by IAS when the pump is confirmed running.*

Pressure and differential pressure alarms are used to be monitored by operators. Any stops of seawater pumps are manual.

6.2 Start-up of thruster sea water system

Below example is starting system port aft. The sequence is similar for the other corners.

- *Open sea side valves to the pump,*
- *Confirm manual inlet sea valve is opened,*
- *Confirm manual valve to Thruster and utilities are opened,*
- *Confirm manual overboard valve is opened,*
- *Open overboard valve,*
- *Start the sea water pump by selecting AUTO mode,*
- *Discharge valves will be automatically open by IAS when the pump is confirmed running.*

Pressure and temperature indication and alarms are used to be monitored by operators.

Any stop of sea water pumps is manual.

6.3 Start-up of sea water service pumps

Below example is starting system in Auto mode:

- *Select which pump to be the duty,*
- *Open sea side valves to both pumps,*
- *Confirm manual inlet sea valves are open,*
- *Verify that manual valves are opened and closed,*
- *Verify manual valves to at least one consumer is open,*
- *Start the duty sea water service pump by selecting AUTO mode,*
- *Discharge valves will be automatically open by IAS when the pump is confirmed running.*

6.4 Start-up of fresh water generator feed pumps

Below example is starting system in Auto mode:

- *Select which pump to be the duty,*
- *Confirm manual inlet sea valves are open,*
- *Start the duty fresh water generator feed pump by selecting AUTO mode,*
- *Suction valves will be automatically open by IAS before the pump is running.*

Continued

6.5 Start-up to Burner Boom/Deluge pump
- Open Burner Boom/Deluge pump discharge valve,
- Confirm manual sea water suction valve and manual Burner Boom/Deluge pump discharge valve re open,
- Confirm manual burner boom inlet valves are open,
- Start the Burner Boom/Deluge pump,
- Suction valves will be automatically open by IAS before the pump is running.

7 Black-out recovery
The automatic sequential start-up after restoring of power is required for the main sea water cooling pumps, thruster sea water cooling pumps and sea water service pumps. On the restart after black out, the pumps which were running prior to black out will start again. The remote operated valves are also fail safe-type to remain as is in case of power failure.

10.5.2 Scope and form of typical system functional description—fresh water generating system

Below text contains example of text contained in a typical System Functional Description[4];

1 Introduction
1.1 General
The control and monitoring functions for the Fresh Water Generator (FW) plants is local automatic system.

Integrated Automation System include Water Generation system monitoring functions and control functions by the fresh water tank inlet valve status and automatic start/stop of fresh water generator feed pumps when the fresh water generator start/stop.

1.2 Classification
The Unit complies with the Det Norske Veritas Offshore Standards with the following class designation;

✠1A1 Column Stabilized Drilling Unit (N), HELDK, E0, DYNPOS AUTRO, POSMOOR ATA.

1.3 Rules and Regulations
DNV-OS-202 Automation, safety, and telecommunication systems, Part 6 Chapter 3 Periodically unattended machinery space
IMO 2009 MODU Code
NMD regulations

1.4 Abbreviations
IAS—Integrated Automation System
FW—Fresh Water
MCC—Motor Control Centre

---

4. Based on Kongsberg project

2 References
P&ID, Fresh Water Generating System
P&ID, Sea Water Cooling System
P&ID, Potable Fresh Water Service System
Fresh Water Generator units, 2 off 50 m³ tones/24h
The process ensures fresh water with salinity levels below 30 ppm NaCl.

4 System description
The potable water from the fresh water generators is produced by evaporation. The water is converted from sea water by vacuum distillation in the single stage fresh water generators. The two sea water feed pumps provide sea water to the fresh water generators. The feed pumps are located in port aft and starboard aft pump room.

4.1 Fresh Water Generators
The Fresh Water Generators are started/stopped manually on Control Boxes installed on their plant skids. Fresh water generator feed pump should start/stop automatically by IAS when IAS receive pump start/stop request from fresh water generator.
   Automatic stop will be performed by IAS if inlet valves to both tanks are closed. Fresh water, brine and sea water flow local gauges are provided.

4.2 Salinity control and dumping
The quality of the produced fresh water is monitored by a salinity control unit. On sensing an excessive salt content, the fresh water flow to the fresh water tanks is automatically stopped. In case of high salinity in the distillate water flow, dumping valves will discharge the produced water to the sea.

5 IAS CONTROL AND MONITORING
5.1 Start and stop
Fresh water generators are started/stopped manually from starter panel.
   Fresh water generator request to IAS to start or stop fresh water feed pump which is dedicated as duty pump. If duty pump fails, the standby pump will be running. The selection of duty/standby pump is via IAS MIMIC.
   Fresh water generator is stopped by IAS when the fresh water tank valve is closed.
   When the fresh water tank valve is closed than fresh water generator is stopped and is cooled down with fresh water. Fresh water generator sends signal to IAS for its feeding pump stop. Additional interlocks between feed pump inlet valve and feed pump are provided.

5.2 Alarm and Monitoring
The following signals from fresh water generators are sent to IAS Alarm and Monitoring system:
– *Running;*
– *Common alarm;*
– *Salinity alarm;*
– *FWG feed pump Start request;*
– *FWG feed pump Stop request;*
– *Man/Auto Mode.*
   IAS sends the following signals to each of the fresh water generators:

Continued

- *FWG Stop request,*
- *FWG Common alarm that includes: Emergency Stop, Shell Low Level Alarm, Boiler Low Level Alarm, Shell Vacuum Alarm, Shell High Level Alarm, Lube Oil Low Pressure Alarm, Lube Oil High Temperature Alarm, Compressor Low Capacity Alarm, Distillate High Conductivity Alarm, Low Chemical Level Alarm, Shell High Pressure Alarm, Compressor Seal Air Low Pressure Alarm and Motor auto start failure Alarm.*

6 Blackout recovery
Manual recovery is provided after blackout.

10.6 Piping and instrumentation diagrams (P&ID)

Piping and Instrumentation Diagrams show the piping and equipment with associated instrumentation and control devices. P&ID diagrams usually contain the processes piping with identifications, and additionally the system equipment such as: pumps, heat exchangers, heaters, compressors, separators, valves and their control, fittings, flowmeters, processes control systems and their instrumentation with input/output identification. Symbols used on P&ID Diagrams are according to ISO Standards. Fig. 10.6.1 shows examples of the symbols commonly used on P&ID Diagrams.

On Piping and Instrumentation Diagrams and system Functional Description documents the following abbreviations are often used:

- ATA—Automatic Thruster Assistance
- BCR—DP Backup Control Room
- CCR—Central Control Room
- DGPS—Differential Global Positioning System
- DP or DPS—Dynamic Positioning System
- EG—Emergency Generator
- ESD—Emergency Shutdown System
- F&G—Fire and Gas System
- FO—Fuel Oil
- FW—Fresh Water
- HMI—Human Machine Interface
- HPU—Hydraulic Power Unit
- HT—High Temperature
- HV—High Voltage
- IAS—Integrated Automation System
- ISS—Integrated Safety System
- LAH—Low Level Alarm
- LAL- High Level Alarm
- LCP Local Control Panel
- LECR—Local Engine Control Room
- LP—Local Panel

- LV—Low Voltage
- MCC—Motor Control Center
- MG—Main Generator
- MRU—Motion Reference Unit
- NDU—Network Distribution Unit
- OS—Operator Station
- PAH—High Pressure Alarm
- PAL—Low Pressure Alarm
- PCV—Pressure Control Valve
- PMS—Power Management System
- PRS—Position Reference System
- QCV—Quick Closing Valve
- RIO—Remote Input/Output module
- ROV—Remote Operated Valve
- SW—Sea Water
- TAL—Low Temperature Alarm
- TAH—High Temperature Alarm
- UPS—Uninterruptible Power Supply
- VFD—Variable Frequency Drive
- VRC—Valve Remote Control

10.6.1   Form of typical piping and instrumentation diagram (P&ID)—fuel oil system

Part of an offshore unit Fuel Oil System Piping and Instrumentation Diagram (P&ID) is presented in Fig. 10.6.2 Fuel oil System P&ID Diagram. On this P&ID drawing are shown Bunker Station Pump, Fuel Oil Transfer Pump, Fuel Oil Supply Pump, two (2) Fuel Oil Bunker Tanks, two (2) Service Tanks and Settling Tank. When offshore Unit bunkering is carried out then oil is routed through manual and remote operated valves into the chosen bunker tank. The filling/suction valves are operated remotely. The Fuel Oil Transfer Pump pumps fuel into Setting Tanks where fuel is separated by gravity. There are two (2) Flow Transmitters for calculation of Fuel Oil consumption from Bunker Station, and also Fuel Oil to Settling Tank. On the P&ID diagram are shown Level Indicators (LI), Pressure Indicators (PI), Pressure Transmitters etc. used for IAS control and monitoring functions.

10.6.2   Form of typical piping and instrumentation (P&ID) diagram—fresh water generator

Fig. 10.6.3 presents a Fresh Water Piping and Instrumentation P&ID Diagram of Fresh Water Generator. The Fresh Water Generator is connected to Sea Water and control air systems. Fresh water is produced by this Generator by evaporation. Start/Stop of the Generator is done manually on a control box installed on

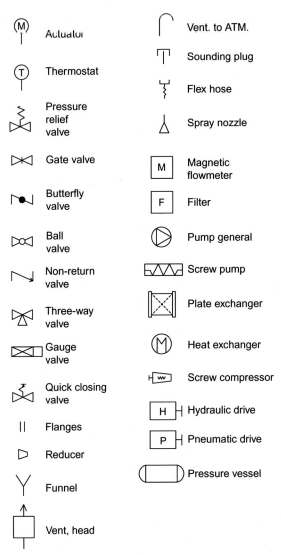

FIG. 10.6.1 Examples of symbols often used on P&ID diagrams. *(Based on ISO 14617 Graphical symbols for diagrams.)*

the Generator Skid. The Fresh Water Generator pump starts/stops automatically when a start/stop request is received from the Generator control box. The quality of the produced water is monitored by a salinity control unit. When the unit senses that salt content is excessive then produced water is discharged to the sea instead of to the fresh water tank.

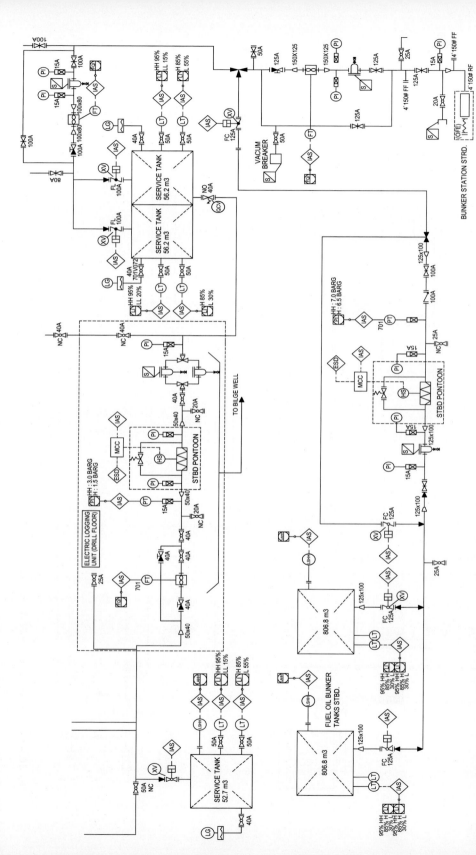

FIG. 10.6.2 Fuel oil system P&ID diagram. (*Based on Kongsberg project.*)

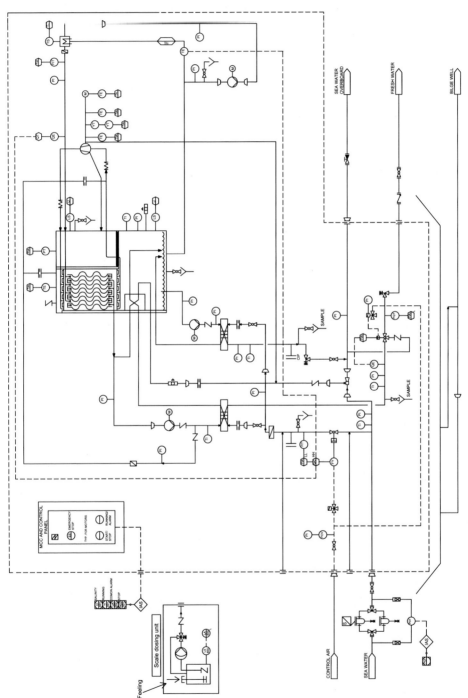

FIG. 10.6.3 Fresh water generator P&ID diagram. *(Based on Kongsberg project.)*

10.6.3   Mimic diagrams

During a project, the Integrated Automation System supplier prepares the Auxiliary Control System SCADA (Supervisory Control & Data Acquisition) from the designed P&IDs. In operation, the crew controls the auxiliary systems using customer defined process mimics presented on the visual display units at the Operator Stations and containing easy to read information on atomised objects. MIMIC Diagram is an arrangement of symbols representing overview of automation process status. The mimic design is developed during the project stage by co-operation between the customer and the IAS supplier. The mimic layout, colours, content, layers and logic should be designed to help the operator understand the process main functions and system layout, and to navigate easily through the different systems.

The systems overview, equipment status (running, stopped) and remotely measured physical data are displayed on custom made processes mimics.

Examples of Kongsberg Mimic Diagrams[5] are presented on Figs. 10.6.4 to 10.6.11.

Usually the following colours are used by Kongsberg to present different systems on Operator Station VDUs:

– Green—Seawater, ballast water, fire water;
– Blue—Fresh water, condensate
– Cyan—Compressed air, CO_2;
– Orange—Lubrication oil;
– Yellow—Hydraulic oil;
– Light Brown—Fuel oil;
– Black—Bilge, sludge, black water, grey water;
– Grey—Steam;
– Brown—Crude oil;
– Green—Electrical System—Single Line Diagram;
– Blue—AC Supply air;
– Light Orange—AC Exhaust air;
– Light Brown—AC Return air.

A typical Offshore Unit Sea water cooling system presented in Fig. 10.6.4 consists of eight (8) main sea water pumps taking water from Sea chests to the central water coolers. Each set of main Sea Water pumps supplies two Central Fresh Water Coolers. Between the inlet and outlet of each Central Water Cooler pipeline is installed a differential pressure indicator which can generate a HIGH differential pressure alarm in the Alarm and Monitoring system. Additionally on the discharge from each Central Cooler discharge is installed a Temperature indicator which can generate HIGH and HIGH HIGH alarms in the control system. Two sea water pumps are installed in each of the four Pump Rooms (Port forward,

---

5. Examples from Kongsberg project

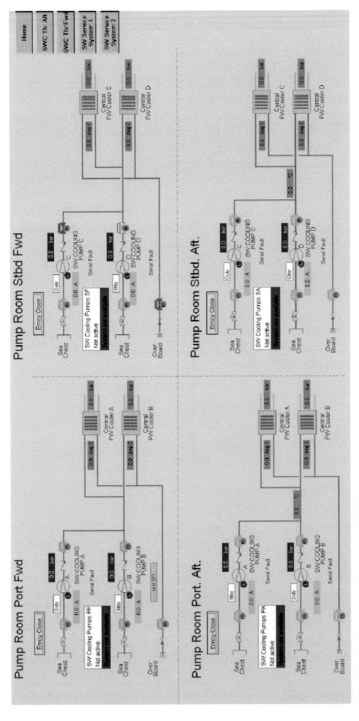

FIG. 10.6.4 Sea water cooling system MIMIC diagram.

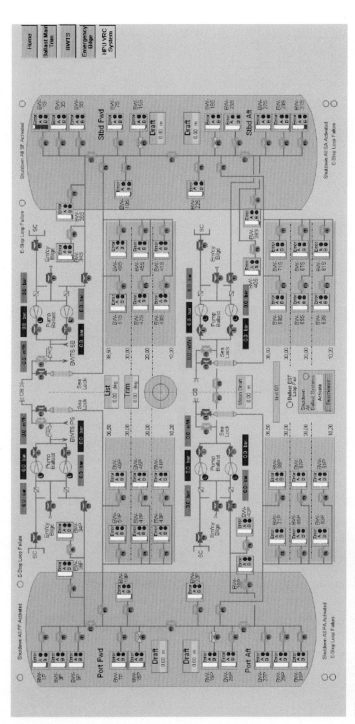

FIG. 10.6.5 Ballast system MIMIC diagram.

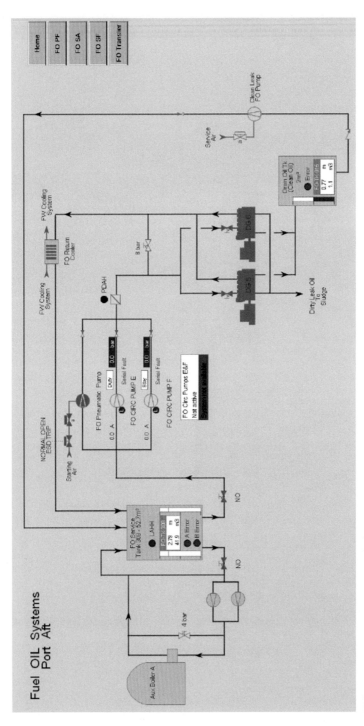

FIG. 10.6.6 Fuel oil system MIMIC diagram.

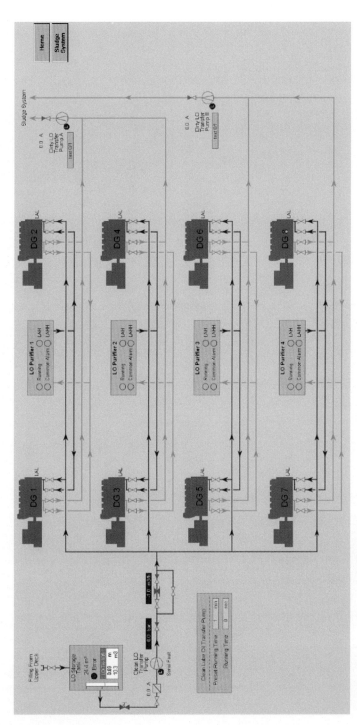

FIG. 10.6.7 Lube oil system MIMIC diagram.

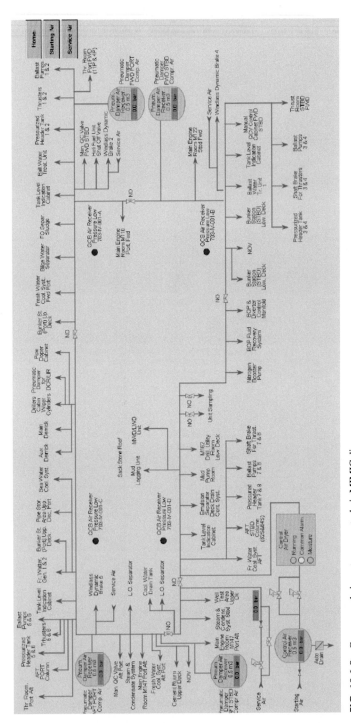

FIG. 10.6.8 Compressed air system—control air MIMIC diagram.

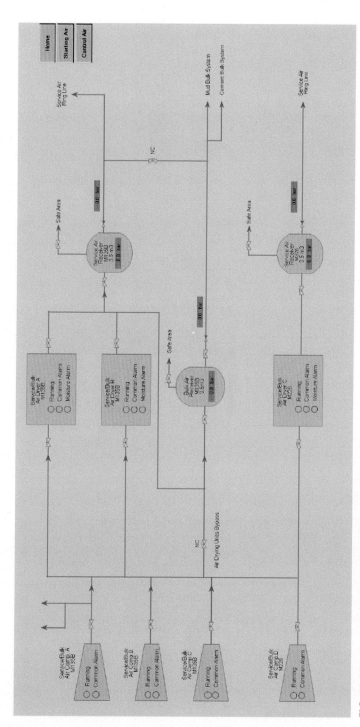

FIG. 10.6.9 Compressed air system—service air MIMIC diagram.

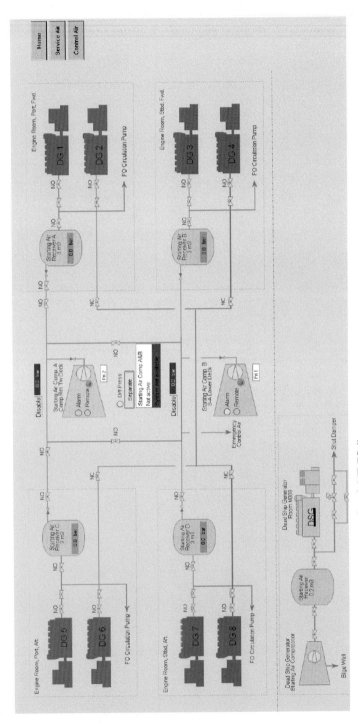

FIG. 10.6.10 Compressed air system—starting air MIMIC diagram.

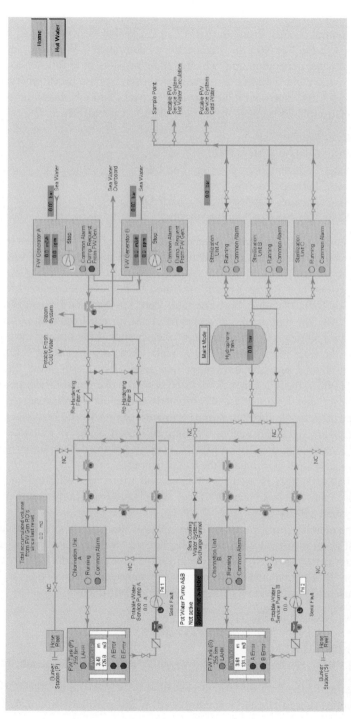

FIG. 10.6.11 Potable fresh water service system circulation MIMIC diagram.

port aft, starboard forward and starboard aft). Seawater is discharged overboard from the central coolers. From this MIMIC it is possible to switch to Sea water service system No 1 or No 2 serving machinery consumers—Windlass, Sewage unit, Fresh water generation unit, winch braking resistors or to thruster cooling system serving eight thrusters and accessories—three phase transformers, frequency converters, electrical motors, hydraulic packs and lubrication units. To make them useful and readable there are available six (6) MIMICS:

- Sea Water Cooling System Aft
- Sea Water Cooling System Fwd.
- Sea Water Thruster Cooling System Port Fwd
- Sea Water Thruster Cooling System Port Aft
- Sea Water Thruster Cooling System Stbd. Aft
- Sea Water Thruster Cooling System Stbd. Fwd

Sea chest valves are operated manually from the IAS system. Sea water discharge valves are normally closed and are opened automatically by the IAS system.

An offshore unit Ballast system presented in Fig. 10.6.5 consists of four (4) pairs of Ballast pumps: Port Forward, Port Aft, Starboard Forward, and Starboard Aft. The pumps are connected by crossover pipelines with normally closed valves so they can act as backup for each other. In ballasting the water passes by the filter to the desired ballast tank. Each ballast pump is equipped with a differential pressure transmitter that gives High alarm when filter is clogged. In de-ballasting water flows directly by the overboard discharge line. Each ballast pump is equipped with a remotely controlled IAS discharge valve. Discharge pressure of all ballast pumps is monitored by IAS and high/low pressure alarms are signalled. Suction pressure of ballast pumps is monitored on IAS and locally. Additionally on MIMIC is presented the differential pressure between suction and discharge of the ballast pumps.

The ballast system is used for offshore unit trimming, ballasting and de-ballasting. Trim, List and Mean Draft indication is presented in the centre of the ballast system MIMIC.

On the pontoon profile are shown the ballast water tanks where the volume of ballast water is presented on bar graphs. In each tank, the water value is calculated as the average of two sensor readings. If any of the sensors is faulty then an "A" or "B" error on the MIMIC bar graph is indicated.

The user may switch from ballast system main MIMIC to:

- Ballast Main Trim MIMIC
- BWTS—Ballast Water Treatment System MIMIC
- Emergency Bilge MIMIC
- HPU VRC—Hydraulic Power Unit Vessel Remote Control System MIMIC.

In design of the Fuel oil system presented in Fig. 10.6.6 the fuel service tank supplies fuel to Diesel Generator No 5, 6 and Auxiliary Boiler A. Fuel

Oil Circulation Pumps take fuel oil from the Fuel Oil Service Tank and discharge the oil to Diesel Generators. Return oil from diesel engines is circulated through the Fuel Oil Return cooler and is discharged back to the service tank. One Circulation Pump is main/working (in service) and the other is in standby mode (starting upon failure of the running pump). There is an additional pneumatic fuel oil pump for working in blackout situations. The Fuel Oil Pumps are controlled remotely from the IAS or locally in case of IAS failure. The working circulation pump is running before the engine is started, and continues to run even if the engine is stopped. The working pump is stopped automatically on High High pressure alarm at discharge of the pump.

The Service Oil tank is equipped with two sensors signalling Low Low, Low, High and High High level alarms. In the Service Oil Tank the oil filling is calculated as the average of two sensor readings. If any of the sensors is faulty then an "A" or "B" error is indicated on the MIMIC bar graph. Clean fuel oil from Diesel Generators No 5 and 6 goes to the Clean Oil tank and is pumped by the Clean Leak Oil Pump to the Fuel Oil Service Tank. In the Drain Oil Tank the level of oil is calculated by sensor. If the sensor is faulty then a sensor alarm is indicated on the MIMIC bar graph.

User may switch from Fuel Oil System Port Aft MIMIC to:

- Fuel oil service port forward MIMIC
- Fuel oil service starboard forward MIMIC
- Fuel oil service starboard aft MIMIC
- Fuel oil transfer MIMIC.

Lubrication Oil system presented in Fig. 10.6.7 consists of LO Storage Tank, Clean LO transfer Pump, four (4) LO Purifiers, eight (8) consumers (Diesel Generators) and two (2) Dirty LO Transfer Pumps. The pressure in the discharge line from LO Storage Tank is monitored by the IAS with Low and High alarms. Additionally the IAS is monitoring the lubrication oil flow to Diesel Generators. Dirty Lubrication oil is transferred by Dirty Oil Transfer Pumps from the diesel engine sumps to the dirty oil/sludge tank. Lubrication oil system valves are operated locally. Lubrication oil separators are started and stopped manually from their local control panels.

The user may switch from Lubrication Oil System MIMIC to Sludge system MIMIC.

In design presented in Fig. 10.6.8 the Compressed Control Air is supplied from Service Air Receiver or from the Starting Air system by a pressure reducer (30 to 7.5 bar), and from the Control Air Receiver and Control Air Dryer. On this offshore unit four (4) pneumatic Damper Receivers are in the system that supply the following air receivers:

- Fresh Water Cooling system
- Sea Water Cooling system
- Fresh Water Generators

- Aft Port Starboard and Fore Port, Starboard Columns
- Pressured Header Tank
- Ballast pumps
- Thrusters
- Shaft brake for Thruster
- Tank Level Indication cabinets
- Emulsion Separator Deck Drain Control system
- Unit sampling
- Nitrogen Booster Pump
- Fluid recovery system
- Bunker Station
- Ballast Water Treatment Unit
- Windlass Dynamic Brake
- Fuel Oil Separator Sludge
- Bilge water Separator etc.

The IAS monitors control air line pressure with Low alarm signalisation, and the Air Dryer common alarm and moisture signalisation.

User may switch from the Compressed Air—Control MIMIC to:

- Compressed Air—Starting Air MIMIC
- Compressed Air—Service Air MIMIC.

In design presented in Fig. 10.6.9 the Compressed Air—Service is supplied from four (4) fresh water cooled screw compressors. The compressed air goes to three (3) air dryers and then to the service air receivers. Each compressor is controlled locally from its Local Control Panel (LCP). The IAS receives signals when the compressor is running and in case of air compressor failure. These signals are presented on Compressed Air—Service Air MIMIC. In normal operation the Service Air compressors are set on their Local Control Panels (LCP) in Auto mode. One machine is Lead, two are Lag and the fourth compressor is in standby mode. The compressors are controlled by pressure transmitter. Low pressure signal on pressure transmitter starts the compressor set in Lead mode. The compressor runs until air receiver pressure reaches set operational pressure. Low, Low Low and High air pressure alarms are signalled on the compressor Local Panel (LCP). Compressed air dryers running status and moisture on their outlets alarm are signalled on the IAS MIMIC.

User may switch from Compressed Air—Service Air MIMIC to:

- Compressed Air—Starting Air MIMIC
- Compressed Air—Control Air MIMIC.

The Starting Air System presented in Fig. 10.6.10 is supplied from two (2) Starting Air Compressors remotely controlled from IAS with means for lead/lag mode operation. Compressors start/stop is based on the air pressure in the system. Additionally Starting Air Compressors are controlled from their Local

Control panels (LCP). Low pressure in any of starting air transmitters automatically starts the compressor set in operational Lead mode.

There are four signals from compressor to IAS: Remote control—status of switch in LCP Local or Remote, Compressor, Compressor running—i.e. compressor motor is running, Compressor common trip—i.e. Local Control Panel (LCP) common alarm/compressor is tripped, and Compressor shut down. There are two signals from compressor to IAS: Start/Stop—i.e. starting and stopping compressor, and Load/Unload.

There are four (4) Air receivers, each sized to start two (2) Generator Main Engines. Each starting air receiver has a pressure transmitter connected to IAS. Two additional transmitters are mounted on the starting air pipelines.

The Starting Air system includes additional Dead Ship Generator Starting Air compressor and additional Starting Air receiver for Dead Ship Generator.

User may switch from Compressed Air—Starting Air MIMIC to:

– Compressed Air—Service Air MIMIC
– Compressed Air—Control Air MIMIC.

Potable Fresh Water presented in Fig. 10.6.11 system consists of two (2) storage tanks, two (2) service pumps, two chlorination units, Hydrophore tank, two (2) fresh water generators and three (3) sterilization units and two (2) re-hardening filters. Fresh water is bunkered via two (2) bunker stations or is produced by two (2) fresh water generators.

Potable water is produced by water generators by evaporation and the distilled water is treated in the re-hardening filter units where minerals are added to produce drinking water quality and is discharged via chlorination units to storage tanks. Bunkered water is also transferred through chlorination units.

Potable water service pumps take water from storage tanks to the hydrophore tank. Discharged water is sterilized in the UV-sterilizing units and is transferred to the cold or hot water distribution system.

Service pumps are operated in Manual or Auto mode. One of the pumps is Duty and the other is Stand-by. In Auto mode the pumps are controlled by IAS that receives pressure value from pressure transducer. Filling valves, recirculation valves, pump suction valves and pump discharge valves are controlled remotely from IAS. Additionally the IAS monitors Hydrophore tank pressure Low/High, Sterilization unit discharge pressure High/Low, Tank Fresh water level High High/High/Low and Sterilizer running, Sterilizer common alarm are signalized.

User may switch from Potable Fresh Water System MIMIC to Hot Water MIMIC.

Examples of Wartsila MIMIC Diagrams

A typical Wartsila MIMIC diagram is presented in Fig. 10.6.12.

In the top information bar main Menu and Sub Menus are displayed. On the bottom information bar are displayed the Login/status, supplemented with an

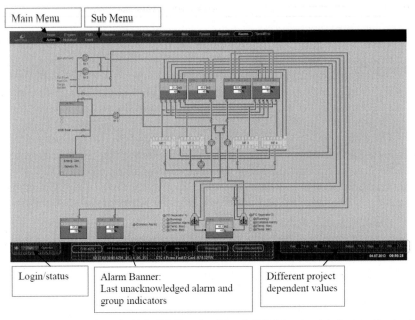

FIG. 10.6.12 Wartsila typical MIMIC. *(Based on Wartsila WIAS functional design specification.)*

Alarm Banners such as last unacknowledged alarm and group indicators. On the right hand side of the bottom information bar are different project values.

The colours normally used by Wartsila to present different information on Visual Display Units (VDUs) are shown in Fig. 10.6.13 Colour definition for Processes and Electrical Systems and Fig. 10.6.14 Colour Definition for Animated Objects and Indicators.

Examples of Wartsila Integrated Automation System (WIAS) MIMIC drawings presented on Operator Station monitors[6] are given in Figs. 10.6.15 to 10.6.18.

Presented in Fig. 10.7.15 is a Lubricating Oil (LO) system consisting of four (4) oil Separators, with four (4) tanks and the associated pumps and heaters supplying four (4) engines. Waste oil from the Oil Separators is transferred to a Waste Oil Tank and is provided with high/low oil level signalization, and tank level measurement in terms of oil volume and tank filling percentage. Additionally LO temperature after the heater is displayed on the MIMIC. On the LO separator icon are provided the following signals: Separator running, Separator common alarm and Max or Min LO temperature alarm.

Presented in Fig. 10.7.16 a Main Engine No 1 Exhaust Gas system is shown on the MIMIC in a black color. On the MIMIC is shown the Exhaust Gas temperature in each of the eight cylinders, cylinder average temperature and Exhaust Gas deviation from average temperature for each cylinder. Additionally, the

---

6. Based on Wartsila documentation

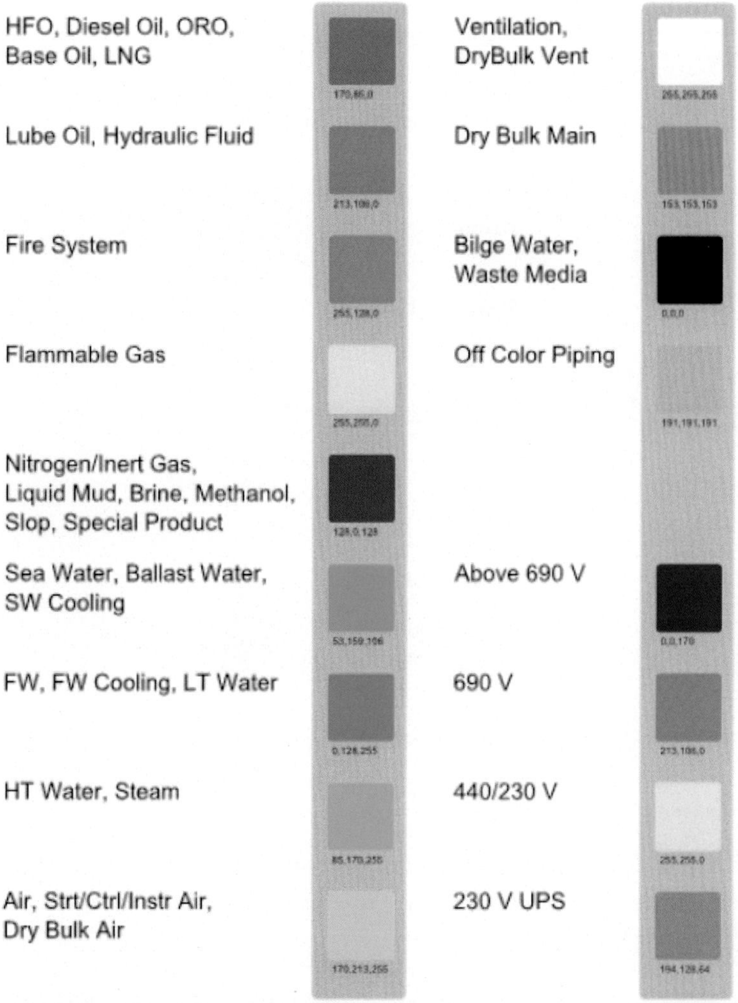

HFO, Diesel Oil, ORO, Base Oil, LNG	170,85,0
Lube Oil, Hydraulic Fluid	213,106,0
Fire System	255,128,0
Flammable Gas	255,255,0
Nitrogen/Inert Gas, Liquid Mud, Brine, Methanol, Slop, Special Product	128,0,128
Sea Water, Ballast Water, SW Cooling	53,159,106
FW, FW Cooling, LT Water	0,128,255
HT Water, Steam	85,170,255
Air, Strt/Ctrl/Instr Air, Dry Bulk Air	170,213,255

Ventilation, DryBulk Vent	255,255,255
Dry Bulk Main	153,153,153
Bilge Water, Waste Media	0,0,0
Off Color Piping	191,191,191
Above 690 V	0,0,170
690 V	213,106,0
440/230 V	255,255,0
230 V UPS	194,128,64

FIG. 10.6.13   Colour definition for processes and electrical systems. *(Based on Wartsila WIAS functional design specification.)*

MIMIC indicates Exhaust Gas temperature before the Exhaust Gas fan and before the Exhaust Gas silencer.

Presented in Fig. 10.6.17 is presented a Fresh Water Cooling System for Main Engine (ME) No. 3 and 4. The system consists of fresh water supply to ME Marine Diesel Oil (MDO) Coolers, Generator Cooler, Preheater and Glycol Heater. In this design each ME Fresh Water cooling system has its own Expansion tank, and each Generator has a separate Box Cooler(s).

Presented in Fig. 10.6.18 is presented a Fuel Oil System supplying Main Engines (ME), Emergency Generator (EG) and an Incinerator. In this design

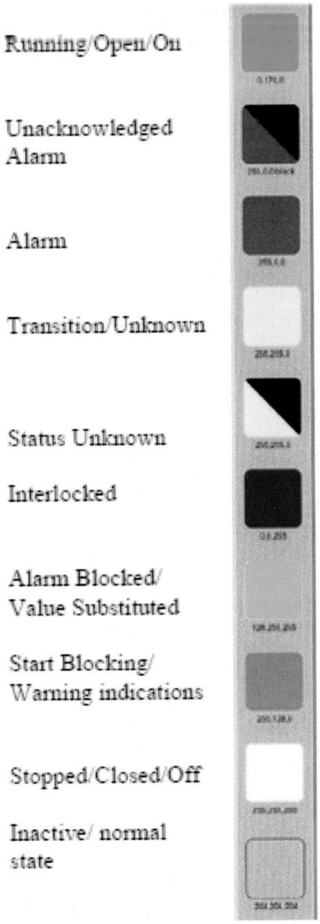

FIG. 10.6.14 Colour definition for animated objects and indicators. *(Based on Wartsila WIAS functional design specification.)*

each engine is supplied by a main or standby MDO pump. The Main Engines are supplied from six (6) Marine Diesel Oil (MDO) tanks. The MIMIC system provides high/low fuel oil level signalization, fuel oil tank volume tank filling percentage. There are two (2) overflow fuel oil tanks, forward and aft provided with high level signalization. The emergency generator has its own tank provided with low level signalization. An additional tank is provided for the fuel oil Incinerator. The MIMIC provides signalization for Incinerator running and Incinerator alarm. The system also includes one Fuel Oil Separator and a Sludge Tank. The MIMIC provides indication of Fuel Oil Separator Running, Common Alarm, Max. Fuel Oil temperature and Min. Fuel Oil temperature, and sludge tank volume and tank filling percentage measurement.

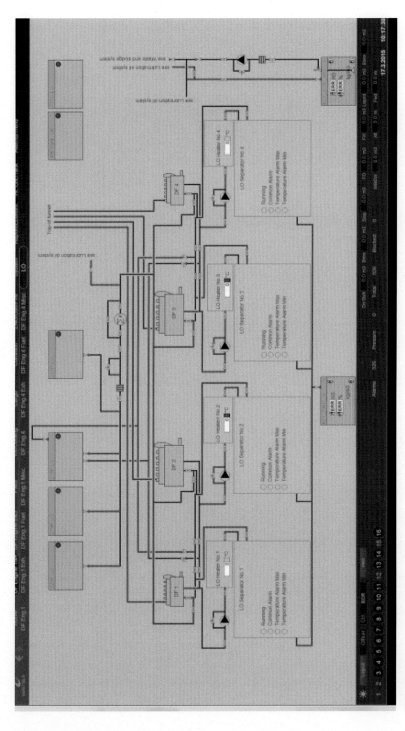

FIG. 10.6.15 Lube oil system.

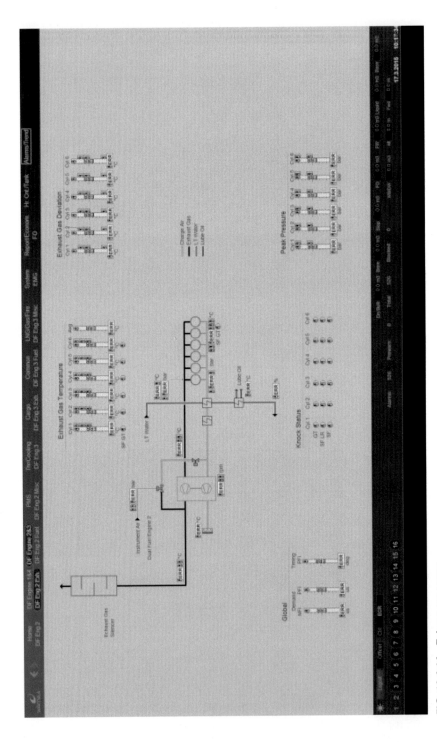

FIG. 10.6.16 Exhaust gas system.

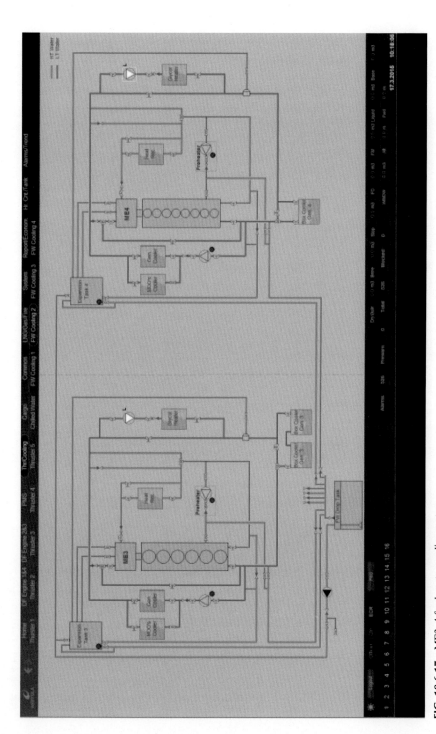

FIG. 10.6.17 ME3_4 fresh water cooling system.

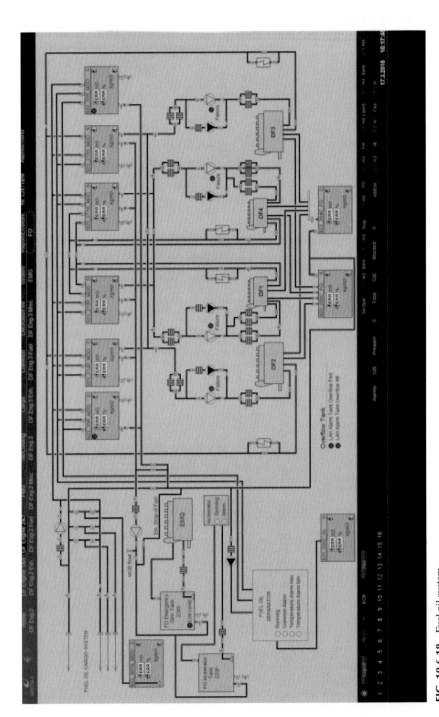

FIG. 10.6.18 Fuel oil system.

Chapter 11

Integrated Automation System (IAS)

Chapter outline

Abstract

This chapter describe typical vessel or offshore Unit Integrated Automation Systems with associated subsystems: Auxiliary Control System, Alarm and Monitoring System and Power Management System, and their interconnections. Additionally this chapter includes example of Integrated Control and Monitoring System (ICMS) topology, a typical Alarm and Monitoring list, print screen of Alarms in Bridge Operator Station and Main Power Distribution System Single Line Diagram (SLD).

11.1 Kongsberg Integrated Automation System (IAS)

Sophisticated computer based Integrated Automation Systems (IAS) provide supervisory control, alarm and data acquisition functions and integrate the control and monitoring systems into one single system.

Integrated Automation System Bridge on *Safe Notos* Accommodation Rig is presented in Fig. 11.1.1.

IAS systems are usually based on Distributed Processing Units (DPU) in which the various process parameters are controlled via I/O Field Stations (FS), located in different parts of the vessel.

Ship and Mobile Offshore Unit Automation. https://doi.org/10.1016/B978-0-12-818723-4.00011-9

FIG. 11.1.1 *Safe Notos* Accommodation Rig Bridge. *(Picture received from Prosafe AS, Norway.* © *David Styles.)*

MMI—Man Machine Interfaces for the IAS are provided by way of multiple redundant Operator Stations (O/S).

The Operator Stations provide an operator interface to the vessel or offshore unit control system and process, the alarm list, trends and reporting system. It is possible to change limits or parameters on one Operator Station and that information is then automatically updated in all relevant Operator Stations.

Access to equipment control and change of process value limits is protected by an Access Control System (ACS) which includes user login passwords assigned to different user groups and functions.

All monitoring and automation functions are carried out by the Distributed Processing Units. Communication between Operator Stations is via redundant Local Area Network (LAN), while communication between Operator Stations and Field Stations is via redundant Controller Area Network (CAN).

An example of Integrated Automation System topology is shown in Fig. 11.1.2.

An Integrated Automation Systems (IAS) usually includes different sub-systems:

- Auxiliary Control System,
- Alarm and Monitoring System,
- Power Management System.

When the IAS is additionally integrated with Safety Management Systems such as Fire & Gas and ESD, then the overall system can be called Integrated Control and Safety System (ICSS) or Integrated Control and Monitoring System (ICMS).

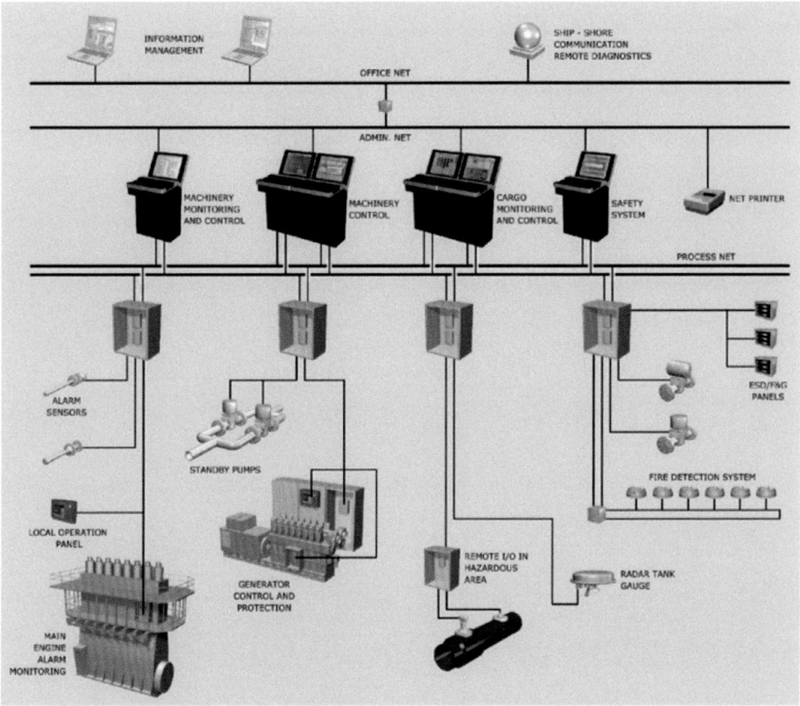

FIG. 11.1.2 Integrated Automation System (IAS). *(Brochure on Kongsberg webpage https://www.km.kongsberg.com/ks/web/nokbg0240.nsf/AllWeb/B34BE0784AD54E48C12572660040AAFA?OpenDocument.)*

Example of ship Integrated Control and Monitoring System (ICMS) topology in presented in Fig. 11.1.3.

11.1.1 Auxiliary control system

The main function of the Auxiliary Control System is to provide a convenient user control interface and ensure that onboard auxiliary systems operate safety and efficiently within design constraints and alarm limits. Typical Auxiliary Control System modules include pump and valve control, tank level gauging, PID controllers etc.

The Auxiliary Control System is part of the IAS and monitors and controls various auxiliary systems. Each control system is designed by Auxiliary Control System designer/supplier according to the overall vessel or offshore unit Technical Specification and systems Functional Description (FD) or Systems Philosophies (SP). Owners and designer's requirements are represented by completed Piping and Instrumentation Diagrams (P&ID) and control and monitoring Input/Output (I/O) Lists.

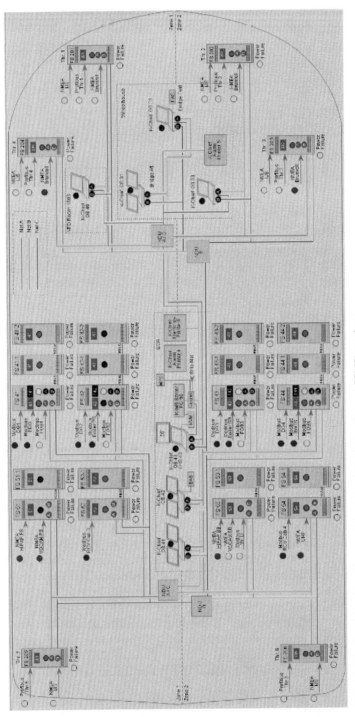

FIG. 11.1.3 Example of ship Integrated Control and Monitoring System (ICMS) topology.

Onboard auxiliary equipment is controlled by the operator using Operator Stations (OP) installed in the Engine Control Room (ECR), on the Navigation Bridge and in DP Stations (for Offshore Units). Each Operator Station is a marine grade computer with colour display and operator panel including trackball.

11.1.2 Alarm and monitoring system

The alarm and monitoring systems provide relevant status and performance information of all machinery systems onboard, and gives audible and visual indication of any faults that requires crew attention. The system is equipped with built-in self-diagnostic properties because of its influence on ship/offshore unit safety. The main functions of the alarm and monitoring system are data processing to initiate alarm, indication of data value and data storage.

The system transfers data to the Bridge, the duty engineers and to the public areas by Watch Calling Modules (WCM). The alarm system indicates vessel equipment abnormal condition, while the monitoring system provides sufficient information for safe operation and to initiate necessary actions. The alarm and monitoring system often includes reporting functions stored in the system and printed when necessary.

New Blackout alarm is presented in Fig. 11.1.4. On the MIMIC diagram are four (4) 690V busbars with connected Generators. It can be seen that at times 12:30:58, 12.10.2018 was initiated alarm with Tag 611.XA.100.01 described MSB1 Busbar 1—Deadbus from Field Station FS41.

Examples of alarms from the alarm and monitoring systems include; "limit values of parameters have been exceeded", "safety system has operated", "failure of power supply to automatic system", or "stand-by power supply has been switched on".

The required scope of alarm and monitoring parameters is defined by the Class societies in their classification rules. In practical execution, the supplier of the alarm and monitoring system receives from the shipyard or owner the agreed list of alarms and parameters to be monitored, and the locations where alarms are to be displayed.

Example of one page of a typical Alarm and Monitoring list is shown in Fig. 11.1.5.

Alarms are announced by visual indication on Operation Station (OS) Visual Display Units (VDU) and on Engine Room Signalling panels. Status of each alarm is distinguished between active alarm, unacknowledged, acknowledges and blocked. Silencing the alarm does not change the alarm indication signal. Acknowledgement of the alarm visual indication changes visual indication and may silence the audible alarm if not silenced before. Active alarms do not prevent indication of any new alarms.

An example of alarms in a Bridge Operator Station (OS) generated by the Alarm and Monitoring system is shown in Fig. 11.1.6.

On the Operator Station' (OP) Video Display Units (VDU) alarms are presented on Dynamic alarm page, Static alarm page and Dynamic event page.

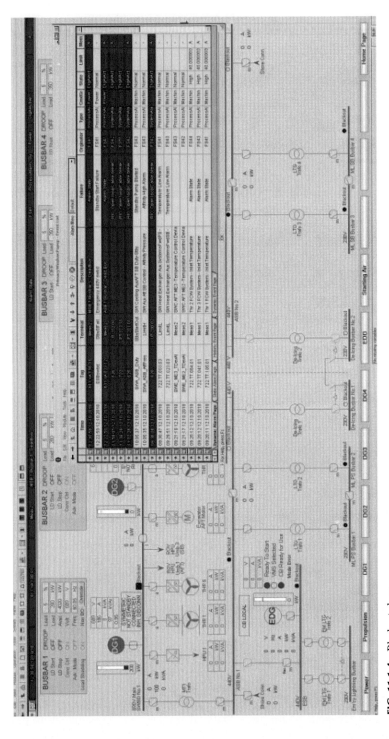

FIG. 11.1.4 Blackout alarm.

Serial No	Description	Sensor			Alarm			ECR			Remark
		Type	Signal	Setting	Print	Block	Delay [s]	CRT Disp.	Meter	Extention	
300	ME Lub Oil System										
301	Main Lub Oil Inl. Press. Low	PT108	4-20mA	0.17MPa	x	a	6-8s	x	x	c	
302	Main Lub Oil Inl. Temp. High	TE112	PT100	55°C	x		6-8s	x	x	c	
303	Piston Cooling Oil Int. Press. Low	PT113	4-20mA	0.17MPa	x	a	6-8s	x	x	c	
304	No 1 Piston Cooling Oil Out Temp. High	TE113.1	PT100	70°C	x		6-8s	x	x	c	
305	No 2 Piston Cooling Oil Out Temp. High	TE113.2	PT100	70°C	x		6-8s	x	x	c	
306	No 3 Piston Cooling Oil Out Temp. High	TE113.3	PT100	70°C	x		6-8s	x	x	c	
307	No 4 Piston Cooling Oil Out Temp. High	TE113.4	PT100	70°C	x		6-8s	x	x	c	
308	No 5 Piston Cooling Oil Out Temp. High	TE113.5	PT100	70°C	x		6-8s	x	x	c	
309	No 6 Piston Cooling Oil Out Temp. High	TE113.6	PT100	70°C	x		6-8s	x	x	c	
310	No 1 Piston Cooling Oil Out Non-Flow	FS114.1	NC	no-flow	x	a	6-8s	x	x	c	
311	No 2 Piston Cooling Oil Out Non-Flow	FS114.2	NC	no-flow	x	a	6-8s	x	x	c	
312	No 3 Piston Cooling Oil Out Non-Flow	FS114.3	NC	no-flow	x	a	6-8s	x	x	c	
313	No 4 Piston Cooling Oil Out Non-Flow	FS114.4	NC	no-flow	x	a	6-8s	x	x	c	
314	No 5 Piston Cooling Oil Out Non-Flow	FS114.5	NC	no-flow	x	a	6-8s	x	x	c	
315	No 6 Piston Cooling Oil Out Non-Flow	FS114.6	NC	no-flow	x	a	6-8s	x	x	c	
316	Thrust Pad Temp High	TE107	PT100	75°C	x		6-8s	x	x	c	
317	No 1 T/C Lub Oil Inlet Press. Low	PT103.1	4-20mA	0.12mPa	x	a	6-8s	x	x	c	
318	No 2 T/C Lub Oil Inlet Press. Low	PT103.2	4-20mA	0.12mPa	x	a	6-8s	x	x	c	
319	No 1 T/C Lub Oil Outlet Temp. High	PT117.1	PT100	95°C	x		6-8s	x	x	c	
320	No 2 T/C Lub Oil Inlet Press. Low	PT117.2	PT100	95°C	x		6-8s	x	x	c	
321	Oil Mist in Crankcase Density High		NC	High	x		6-8s	x	x	c	
322	Oil Mist Detector Failure		NC		x		6-8s	x	x	c	
323	M/E Cyl. L.O. Temp. High	TE202	PT100	70°C	x		6-8s	x	x	c	
324	Leakage Oil From Hyd. Cyl. L.O. Unit High	LS112	NC	High	x		6-8s	x	x	c	
325	M/E Backflush Diff. Press. Too High		NC	0.08mPa	x			x	x	c	
326	M/E Bear Wear System Failure		NC		x			x	x	c	
327	M/E Lub. Oil Water Activity High		NO	50%RH	x		6-8s	x	x	c	

FIG. 11.1.5 Extract from Typical Alarm and Monitoring List.

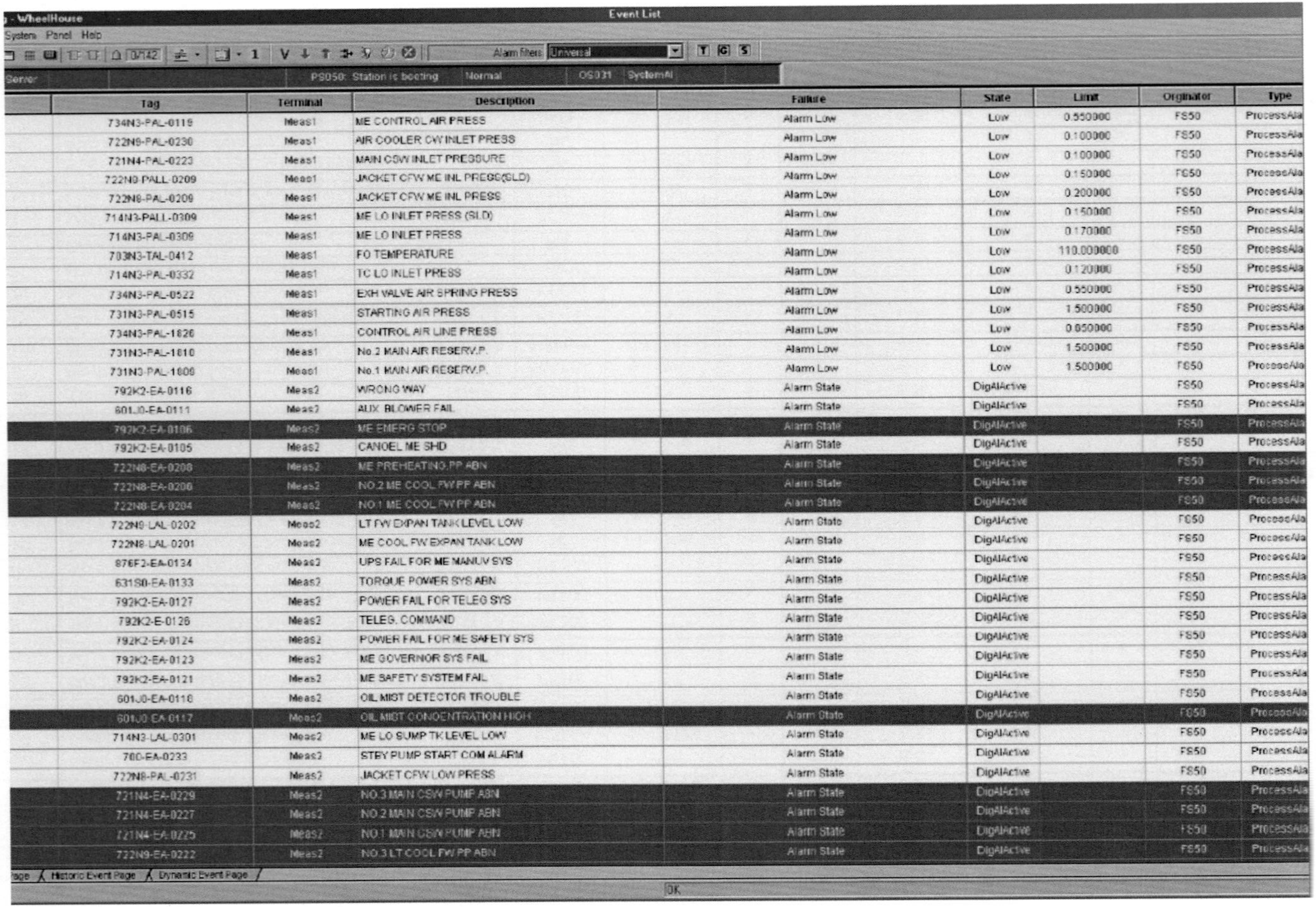

WheelHouse
System Panel Help

Event List

PS050   Stationic booking   Normal   OS331   System Al   Alarm Res   Unread

Tag	Terminal	Description	Failure	State	Limit	Originator	Type
734N3-PAL-0119	Meas1	ME CONTROL AIR PRESS	Alarm Low	Low	0.550000	FS50	Process Ala
722N6-PAL-0230	Meas1	AIR COOLER CW INLET PRESS	Alarm Low	Low	0.100000	FS50	Process Ala
721N4-PAL-0223	Meas1	MAIN COW INLET PRESSURE	Alarm Low	Low	0.100000	FS50	Process Ala
722N0-PALL 0209	Meas1	JACKET CFW ME INL PRESS(SLD)	Alarm Low	Low	0.150000	FS50	Process Ala
722N6-PAL-0209	Meas1	JACKET CFW ME INL PRESS	Alarm Low	Low	0.200000	FS50	Process Ala
714N3-PALL-0309	Meas1	ME LO INLET PRESS (SLD)	Alarm Low	Low	0.150000	FS50	Process Ala
714N3-PAL-0309	Meas1	ME LO INLET PRESS	Alarm Low	Low	0.170000	FS50	Process Ala
703N3-TAL-0412	Meas1	FO TEMPERATURE	Alarm Low	Low	110.000000	FS50	Process Ala
714N3-PAL-0352	Meas1	1C LO INLET PRESS	Alarm Low	Low	0.120000	FS50	Process Ala
734N3-PAL-0522	Meas1	EXH VALVE AIR SPRING PRESS	Alarm Low	Low	0.550000	FS50	Process Ala
731N3-PAL-0515	Meas1	STARTING AIR PRESS	Alarm Low	Low	1.500000	FS50	Process Ala
734N3-PAL-1820	Meas1	CONTROL AIR LINE PRESS	Alarm Low	Low	0.650000	FS50	Process Ala
731N3-PAL-1610	Meas1	No.2 MAIN AIR RESERV.P	Alarm Low	Low	1.500000	FS50	Process Ala
731N3-PAL-1609	Meas1	No.1 MAIN AIR RESERV.P	Alarm Low	Low	1.500000	FS50	Process Ala
792x2-EA-0116	Meas2	WRONG WAY	Alarm State	DigitActive		FS50	Process Ala
801.L0-EA-0111	Meas2	AUX. BLOWER FAIL	Alarm State	DigitActive		FS50	Process Ala
792K2-EA-0106	Meas2	ME EMERG STOP	Alarm State	DigitActive		FS50	Process Ala
792K2-EA-0105	Meas2	CANCEL ME SHD	Alarm State	DigitActive		FS50	Process Ala
722N6-EA-0208	Meas2	ME PREHEATING PP ABN	Alarm State	DigitActive		FS50	Process Ala
722N8-EA-0206	Meas2	No.2 ME COOL FW FP ABN	Alarm State	DigitActive		FS50	Process Ala
722N6-EA-0204	Meas2	No.1 ME COOL FW FP ABN	Alarm State	DigitActive		FS50	Process Ala
722N6-LAL-0202	Meas2	LT FW EXPAN TANK LEVEL LOW	Alarm State	DigitActive		FS50	Process Ala
722N6-LAL-0201	Meas2	ME COOL FW EXPAN TANK LOW	Alarm State	DigitActive		FS50	Process Ala
876F2-EA-0134	Meas2	UPS FAIL FOR ME MANUV SYS	Alarm State	DigitActive		FS50	Process Ala
631S0-EA-0133	Meas2	TORQUE POWER SYS ABN	Alarm State	DigitActive		FS50	Process Ala
792K2-EA-0127	Meas2	POWER FAIL FOR TELEG SYS	Alarm State	DigitActive		FS50	Process Ala
792K2-E-0126	Meas2	TELEG COMMAND	Alarm State	DigitActive		FS50	Process Ala
792K2-EA-0124	Meas2	POWER FAIL FOR ME SAFETY SYS	Alarm State	DigitActive		FS50	Process Ala
792K2-EA-0123	Meas2	ME GOVERNOR SYS FAIL	Alarm State	DigitActive		FS50	Process Ala
792K2-EA-0121	Meas2	ME SAFETY SYSTEM FAIL	Alarm State	DigitActive		FS50	Process Ala
722N6-PAL-0231	Meas2	JACKET CFW LOW PRESS	Alarm State	DigitActive		FS50	Process Ala
601.L0-EA-0116	Meas2	OIL MIST DETECTOR TROUBLE	Alarm State	DigitActive		FS50	Process Ala
601.L0 CA 0117	Meas2	OIL MIST CONCENTRATION HIGH	Alarm State	DigitActive		FS50	Process Ala
714N3-LAL-0301	Meas2	ME LO SUMP TK LEVEL LOW	Alarm State	DigitActive		FS50	Process Ala
760-EA-0233	Meas2	STBY PUMP START COM ALARM	Alarm State	DigitActive		FS50	Process Ala
722N6-PAL-0231	Meas2	JACKET CFW LOW PRESS	Alarm State	DigitActive		FS50	Process Ala
721N4-EA-0229	Meas2	NO.3 MAIN CSW PUMP ABN	Alarm State	DigitActive		FS50	Process Ala
721N4-EA-0227	Meas2	NO.2 MAIN CSW PUMP ABN	Alarm State	DigitActive		FS50	Process Ala
721N4-EA-0225	Meas2	NO.1 MAIN CSW PUMP ABN	Alarm State	DigitActive		FS50	Process Ala
722N9-EA-0222	Meas2	NO.3 LT COOL FW PP ABN	Alarm State	DigitActive		FS50	Process Ala

Page A   Historic Event Page   Dynamic Event Page

OK

FIG. 11.1.6 Example of Alarms on Bridge Operator Station (OS). *(Print screen from Kongsberg Wheelhouse Operator Station VDU.)*

Each page includes Alarm time, Alarm tag, Alarm terminal, Alarm and Failure description, originator—Field Station number, Command group, Alarm state and member.

Examples of these alarm pages are presented in the Figs. 11.1.7, 11.1.8 and 11.1.9. Yellow highlighted alarms are low priority and Red highlighted alarms are high priority alarms.

11.1.3 Power Management Systems (PMS)

The Power Management System (PMS) is often provided as part of the IAS and provides control of electrical generators, switchboards and large consumers. The primary function of the Power Management System is to ensure that power capacity is in line with vessel power demand at any time. The PMS ensures that the load from main consumers does not overload power plant capacity, even if one of the generators should shut down unexpectedly. The PMS will automatically start-up and stop spare generators when required, and may sometimes shed load from large consumers to avoid overload.

A power management system is usually performing the following functions:

- Generator set control and monitoring—Generator can be connected when engine is started and generator voltage is established. 'Connect' command triggers PMS to activate generator synchronising process to adjust engine speed, generator voltage, frequency and phases check before closing the circuit breaker. Generator and switchboard synchronising units are usually installed in switchboard generator panels.
- Load dependent start/stop—Number of generators connected to the main switchboards is automatically controlled depending on actual electric power demand.

Load dependent start/stop delay time can be arranged as shown in Fig. 11.1.10.

- Blackout restart—In case of blackout, the first stand-by generator will automatically be started. To prevent generator overload, relevant consumers are reconnected in a pre-determined sequence.
- Load sharing—Basing on active power measurement, the PMS allows for selection of the load sharing mode for individual generators. Options include symmetric load sharing, asymmetric load sharing, and manual load sharing of fixed load.
- Start blocking of heavy consumers—To prevent overloading of generator plant the PMS may restrict start of heavy consumers, for example those with power exceeding 500 kW, unless available power is sufficient and relevant number of generators is connected to the network. When a Start request signal is initiated the PMS is checking power availability to start heavy consumer electric motors. When the power is available and other machinery starting conditions are fulfilled then an Available signal is given to the IAS.

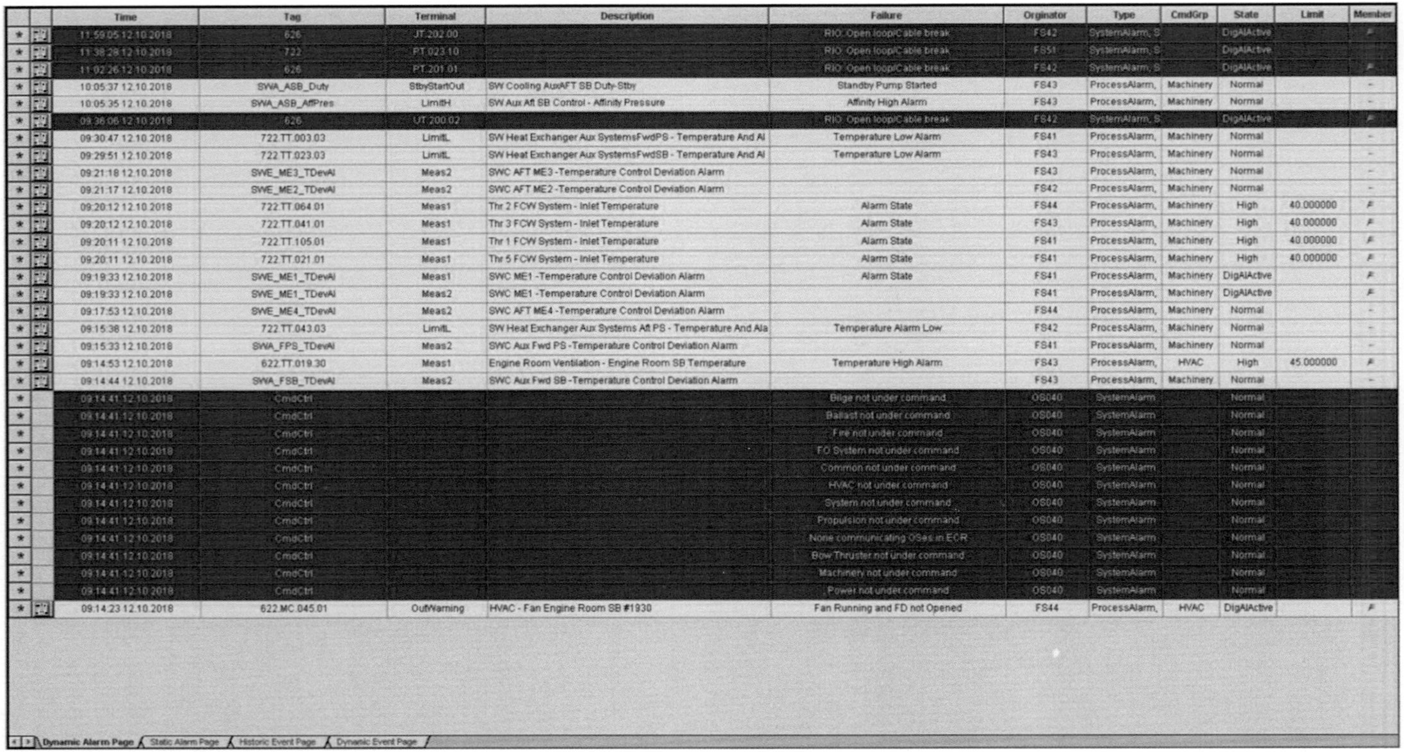

	Time	Tag	Terminal	Description	Failure	Originator	Type	CmdGrp	State	Limit	Members
*	11:59:05 12/10/2018	626	JT.202.00		RIO Open logoiCable break	FS42	SystemAlarm, S		DigitActive		
*	11:38:29 12/10/2018	722	PT.023.10		RIO Open logoiCable break	FS1	SystemAlarm, S		DigitActive		
*	10:26:26 12/10/2018	626	PT.201.01		RIO Open logoiCable break	FS42	SystemAlarm, S		DigitActive		
*	10:05:37 12/10/2018	SWA_ASB_Duty	SbvStartOut	SW Cooling AuxAFT SB Duty-Stbr	Standby Pump Started	FS43	ProcessAlarm,	Machinery	Normal		
*	10:05:35 12/10/2018	SWA_ASB_AtPres	LimBH	SW Aux Aft SB Control - Affinity Pressure	Affinity High Alarm	FS43	ProcessAlarm,	Machinery	Normal		
*	09:36:06 12/10/2018	626	UT.200.02		RIO Open logoiCable break	FS42	SystemAlarm, S		DigitActive		
*	09:30:47 12/10/2018	722.TT.003.03	LimitL	SW Heat Exchanger Aux SystemsFwdPS - Temperature And Al	Temperature Low Alarm	FS41	ProcessAlarm,	Machinery	Normal		
*	09:29:51 12/10/2018	722.TT.023.03	LimitL	SW Heat Exchanger Aux SystemsFwdSB - Temperature And Al	Temperature Low Alarm	FS43	ProcessAlarm,	Machinery	Normal		
*	09:21:18 12/10/2018	SWE_ME2_TDevAl	Meas2	SWC AFT ME3 -Temperature Control Deviation Alarm		FS43	ProcessAlarm,	Machinery	Normal		
*	09:21:17 12/10/2018	SWE_ME2_TDevAl	Meas2	SWC AFT ME2 -Temperature Control Deviation Alarm		FS42	ProcessAlarm,	Machinery	Normal		
*	09:20:12 12/10/2018	722.TT.064.01	Meas1	Thr 2 FCW System - Inlet Temperature	Alarm State	FS44	ProcessAlarm,	Machinery	High	40.000000	
*	09:20:12 12/10/2018	722.TT.041.01	Meas1	Thr 3 FCW System - Inlet Temperature	Alarm State	FS43	ProcessAlarm,	Machinery	High	40.000000	
*	09:20:11 12/10/2018	722.TT.105.01	Meas1	Thr 1 FCW System - Inlet Temperature	Alarm State	FS41	ProcessAlarm,	Machinery	High	40.000000	
*	09:20:11 12/10/2018	722.TT.021.01	Meas1	Thr 5 FCW System - Inlet Temperature	Alarm State	FS41	ProcessAlarm,	Machinery	High	40.000000	
*	09:19:33 12/10/2018	SWE_ME1_TDevAl	Meas1	SWC ME1 - Temperature Control Deviation Alarm		FS41	ProcessAlarm,	Machinery	DigitActive		
*	09:19:33 12/10/2018	SWE_ME1_TDevAl	Meas2	SWC ME1 - Temperature Control Deviation Alarm		FS41	ProcessAlarm,	Machinery	DigitActive		
*	09:17:53 12/10/2018	SWE_ME4_TDevAl	Meas2	SWC AFT ME4 -Temperature Control Deviation Alarm		FS42	ProcessAlarm,	Machinery	Normal		
*	09:15:38 12/10/2018	722.TT.043.03	LimitL	SW Heat Exchanger Aux Systems Aft PS - Temperature And Ala	Temperature Alarm Low	FS42	ProcessAlarm,	Machinery	Normal		
*	09:15:33 12/10/2018	SWA_FPS_TDevAl	Meas2	SWC Aux Fwd PS -Temperature Control Deviation Alarm		FS41	ProcessAlarm,	Machinery	Normal		
*	09:14:53 12/10/2018	622.TT.019.30	Meas1	Engine Room Ventilation - Engine Room SB Temperature	Temperature High Alarm	FS43	ProcessAlarm,	HVAC	High	45.000000	
*	09:14:44 12/10/2018	SWA_FSB_TDevAl	Meas2	SWC Aux Fwd SB -Temperature Control Deviation Alarm		FS43	ProcessAlarm,	Machinery	Normal		
*	09:14:41 12/10/2018	CmdCtrl			Bilge not under command		SystemAlarm	OS040	Normal		
*	09:14:41 12/10/2018	CmdCtrl			Ballast not under command		SystemAlarm	OS040	Normal		
*	09:14:41 12/10/2018	CmdCtrl			Fire not under command		SystemAlarm	OS040	Normal		
*	09:14:41 12/10/2018	CmdCtrl			FO System not under command		SystemAlarm	OS040	Normal		
*	09:14:41 12/10/2018	CmdCtrl			Common not under command		SystemAlarm	OS040	Normal		
*	09:14:41 12/10/2018	CmdCtrl			HVAC not under command		SystemAlarm	OS040	Normal		
*	09:14:41 12/10/2018	CmdCtrl			System not under command		SystemAlarm	OS040	Normal		
*	09:14:41 12/10/2018	CmdCtrl			Propulsion not under command		SystemAlarm	OS040	Normal		
*	09:14:41 12/10/2018	CmdCtrl			Noise communicating OS#1 in ECR		SystemAlarm	OS040	Normal		
*	09:14:41 12/10/2018	CmdCtrl			Bow Thruster not under command		SystemAlarm	OS040	Normal		
*	09:14:41 12/10/2018	CmdCtrl			Machinery not under command		SystemAlarm	OS040	Normal		
*	09:14:41 12/10/2018	CmdCtrl			Power not under command		SystemAlarm	OS040	Normal		
*	09:14:23 12/10/2018	622.MC.045.01	OutWarning	HVAC - Fan Engine Room SB #1930	Fan Running and FD not Opened	FS44	ProcessAlarm,	HVAC	DigitActive		

Dynamic Alarm Page / Static Alarm Page / Historic Event Page / Dynamic Event Page /

FIG. 11.1.7 Dynamic alarm page. (*Print screen from Kongsberg Wheelhouse Operator Station VDU.*)

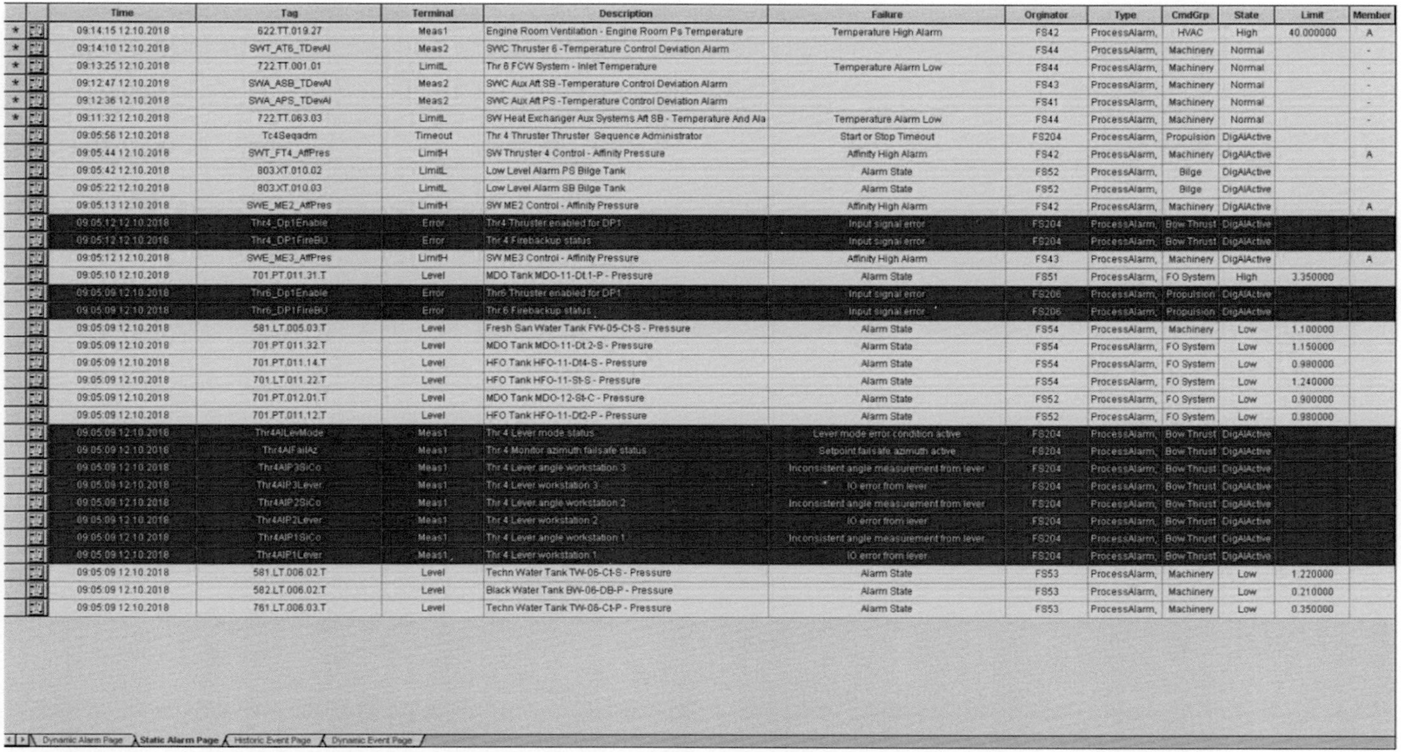

	Time	Tag	Terminal	Description	Failure	Originator	Type	OrsGrp	State	Limit	Member
*	09:14:15 12.10.2018	822.TT.010.27	Meas1	Engine Room Ventilation - Engine Room Fa Temperature	Temperature High Alarm	FS42	ProcessAlarm,	HVAC	High	40.000000	A
*	09:14:10 12.10.2018	SWT_ATB_TDevAl	Meas2	SWC Thruster 6 -Temperature Control Deviation Alarm		FS44	ProcessAlarm,	Machinery	Normal	-	-
*	09:13:25 12.10.2018	722.TT.001.01	LimitL	Thr 6 FCW System- Inlet Temperature	Temperature Alarm Low	FS44	ProcessAlarm,	Machinery	Normal	-	-
*	09:12:47 12.10.2018	SWA_ASB_TDevAl	Meas2	SWC Aux Aft SB -Temperature Control Deviation Alarm		FS43	ProcessAlarm,	Machinery	Normal	-	-
*	09:12:36 12.10.2018	SWA_APS_TDevAl	Meas2	SWC Aux Aft PS -Temperature Control Deviation Alarm		FS41	ProcessAlarm,	Machinery	Normal	-	-
*	09:11:32 12.10.2018	722.TT.063.03	LimitL	SW Heat Exchanger Aux Systems Aft SB - Temperature And Ala	Temperature Alarm Low	FS44	ProcessAlarm,	Machinery	Normal	-	-
	09:05:56 12.10.2018	Tc4Seqadm	Timeout	Thr 4 Thruster Thruster Sequence Administrator	Start or Stop Timeout	FS204	ProcessAlarm,	Propulsion	DigiAActive		
	09:05:44 12.10.2018	SWT_FTL_AffPres	LimitH	SW Thruster 4 Control - Affinity Pressure	Affinity High Alarm	FS42	ProcessAlarm,	Machinery	DigiAActive		A
	09:05:43 12.10.2018	803.XT.010.02	LimitL	Low Level Alarm PS Bilge Tank	Alarm State	FS52	ProcessAlarm,	Bilge	DigiAActive		
	09:05:22 12.10.2018	803.XT.010.03	LimitH	Low Level Alarm SB Bilge Tank	Alarm State	FS52	ProcessAlarm,	Bilge	DigiAActive		
	09:05:13 12.10.2018	SWE_ME2_AffPres	LimitH	SW ME2 Control - Affinity Pressure	Affinity High Alarm	FS42	ProcessAlarm,	Machinery	DigiAActive		A
	09:05:12 12.10.2018	Thr4_Ctr1Enable	Error	Thr4 Thruster enabled for DP1	Input signal error	FS204	ProcessAlarm,	Bow Thrust	DigiAActive		
	09:05:12 12.10.2018	Thr4_DP1FmBu	Error	Thr 4 if feedback status	Input signal error	FS204	ProcessAlarm,	Bow Thrust	DigiAActive		
	09:05:11 12.10.2018	SWE_ME3_AffPres	LimitH	SW ME3 Control - Affinity Pressure	Affinity High Alarm	FS43	ProcessAlarm,	Machinery	DigiAActive		A
	09:05:10 12.10.2018	701.PT.011.31.T	Level	MDO Tank MDO-11-Dt1-P- Pressure	Alarm State	FS51	ProcessAlarm,	FO System	High	3.350000	
	09:05:09 12.10.2018	Thr5_Ctr1Enable	Error	Thr5 Thruster enabled for DP1	Input signal error	FS206	ProcessAlarm,	Propulsion	DigiAActive		
	09:05:09 12.10.2018	Thr6_DP1FmBu	Error	Thr 6 if feedback status	Input signal error	FS206	ProcessAlarm,	Propulsion	DigiAActive		
	09:05:09 12.10.2018	581.LT.006.03.T	Level	Fresh San Water Tank FW-05-Ct-S- Pressure	Alarm State	FS54	ProcessAlarm,	Machinery	Low	1.100000	
	09:05:09 12.10.2018	701.PT.011.32.T	Level	MDO Tank MDO-11-Dt-2-S- Pressure	Alarm State	FS54	ProcessAlarm,	FO System	Low	1.150000	
	09:05:09 12.10.2018	701.PT.011.14.T	Level	HFO Tank HFO-11-Dt4-S- Pressure	Alarm State	FS54	ProcessAlarm,	FO System	Low	0.990000	
	09:05:09 12.10.2018	701.LT.011.22.T	Level	HFO Tank HFO-11-St-S- Pressure	Alarm State	FS52	ProcessAlarm,	FO System	Low	1.240000	
	09:05:09 12.10.2018	701.PT.012.01.T	Level	MDO Tank MDO-12-St-C- Pressure	Alarm State	FS52	ProcessAlarm,	FO System	Low	0.900000	
	09:05:09 12.10.2018	701.PT.011.12.T	Level	HFO Tank HFO-11-DQ-P- Pressure	Alarm State	FS52	ProcessAlarm,	FO System	Low	0.990000	
	09:05:09 12.10.2018	Thr4xLevMode	Meas1	Thr 4 Lever mode status	Lever mode error condition active	FS204	ProcessAlarm,	Bow Thrust	DigiAActive		
	09:05:09 12.10.2018	Thr4AIF_MAC	Meas1	Thr 4 Manual azimuth failsafe status	Setpoint failsafe azimuth active	FS204	ProcessAlarm,	Bow Thrust	DigiAActive		
	09:05:09 12.10.2018	Thr4AIF_PSbCo	Meas1	Thr 4 Lever angle workstation 3	Inconsistent angle measurement from lever	FS204	ProcessAlarm,	Bow Thrust	DigiAActive		
	09:05:09 12.10.2018	Thr4AIF3Lever	Meas1	Thr 4 Lever workstation 3	IO error from lever	FS204	ProcessAlarm,	Bow Thrust	DigiAActive		
	09:05:09 12.10.2018	Thr4AIF2SbCo	Meas1	Thr 4 Lever angle workstation 2	Inconsistent angle measurement from lever	FS204	ProcessAlarm,	Bow Thrust	DigiAActive		
	09:05:09 12.10.2018	Thr4AIF2Lever	Meas1	Thr 4 Lever workstation 2	IO error from lever	FS204	ProcessAlarm,	Bow Thrust	DigiAActive		
	09:05:09 12.10.2018	Thr4AIF1SbCo	Meas1	Thr 4 Lever angle workstation 1	Inconsistent angle measurement from lever	FS204	ProcessAlarm,	Bow Thrust	DigiAActive		
	09:05:09 12.10.2018	Thr4AIF1Lever	Meas1	Thr 4 Lever workstation 1	IO error from lever	FS204	ProcessAlarm,	Bow Thrust	DigiAActive		
	09:05:09 12.10.2018	581.LT.006.02.T	Level	Techn Water Tank TW-05-Ct-S- Pressure	Alarm State	FS53	ProcessAlarm,	Machinery	Low	1.210000	
	09:05:09 12.10.2018	592.LT.006.02.T	Level	Black Water Tank BW-06-Dt-P- Pressure	Alarm State	FS53	ProcessAlarm,	Machinery	Low	0.210000	
	09:05:09 12.10.2018	761.LT.006.03.T	Level	Techn Water Tank TW-06-Ct-P- Pressure	Alarm State	FS53	ProcessAlarm,	Machinery	Low	0.350000	

Dynamic Alarm Page / Static Alarm Page / Historic Event Page / Dynamic Event Page

FIG. 11.1.8   Static alarm page. (*Print screen from Kongsberg Wheelhouse Operator Station VDU.*)

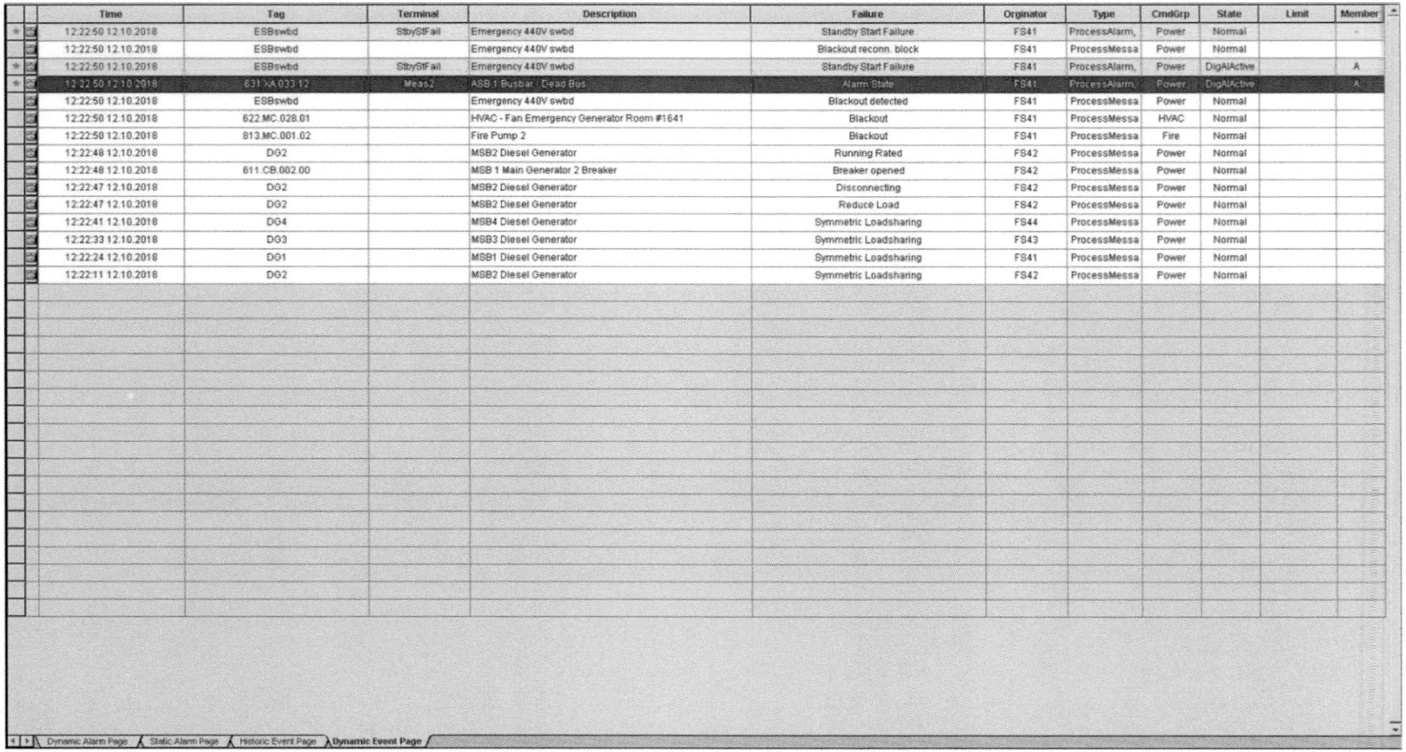

Time	Tag	Terminal	Description	Failure	Orginator	Type	CmdGrp	State	Limit	Member
12:22:50 12.10.2018	ESBswbd	StbyStFail	Emergency 440V swbd	Standby Start Failure	FS41	ProcessAlarm,	Power	Normal		-
12:22:50 12.10.2018	ESBswbd		Emergency 440V swbd	Blackout reconn. block	FS41	ProcessMessa	Power	Normal		
12:22:50 12.10.2018	ESBswbd	StbyStFail	Emergency 440V swbd	Standby Start Failure	FS41	ProcessAlarm,	Power	DigAlActive		A
12:22:50 12.10.2018	631.XA.033.12	Meas2	ASB 1 Busbar - Dead Bus	Alarm State	FS41	ProcessAlarm,	Power	DigAlActive		A
12:22:50 12.10.2018	ESBswbd		Emergency 440V swbd	Blackout detected	FS41	ProcessMessa	Power	Normal		
12:22:50 12.10.2018	622.MC.028.01		HVAC - Fan Emergency Generator Room #1641	Blackout	FS41	ProcessMessa	HVAC	Normal		
12:22:50 12.10.2018	813.MC.001.02		Fire Pump 2	Blackout	FS41	ProcessMessa	Fire	Normal		
12:22:48 12.10.2018	DG2		MSB2 Diesel Generator	Running Rated	FS42	ProcessMessa	Power	Normal		
12:22:48 12.10.2018	611.CB.002.00		MSB 1 Main Generator 2 Breaker	Breaker opened	FS42	ProcessMessa	Power	Normal		
12:22:47 12.10.2018	DG2		MSB2 Diesel Generator	Disconnecting	FS42	ProcessMessa	Power	Normal		
12:22:47 12.10.2018	DG2		MSB2 Diesel Generator	Reduce Load	FS42	ProcessMessa	Power	Normal		
12:22:41 12.10.2018	DG4		MSB4 Diesel Generator	Symmetric Loadsharing	FS44	ProcessMessa	Power	Normal		
12:22:33 12.10.2018	DG3		MSB3 Diesel Generator	Symmetric Loadsharing	FS43	ProcessMessa	Power	Normal		
12:22:24 12.10.2018	DG1		MSB1 Diesel Generator	Symmetric Loadsharing	FS41	ProcessMessa	Power	Normal		
12:22:11 12.10.2018	DG2		MSB2 Diesel Generator	Symmetric Loadsharing	FS42	ProcessMessa	Power	Normal		

FIG. 11.1.9 Dynamic event page. *(Print screen from Kongsberg Wheelhouse Operator Station VDU.)*

Number of DGs connected	Start limit 1	Delay time 1	Start limit 2	Delay time 2	Stop limit	Delay time
1	82%	30 s	88%	10 s	–	–
2	84%	30 s	90%	10 s	65%	15 min
3	86%	30 s	90%	10 s	70%	15 min
4	90%	30 s	100%	10 s	70%	15 min

FIG. 11.1.10 Load dependent generators start/stop delay time. *(Based on Kongsberg project.)*

- Load shedding—To reduce demand on energy generating system, the PMS logic will include switching off less important consumer systems. In the case of dynamically positioned vessels, the PMS may sometimes send a fast load reduction signal to the thrusters so that overall blackout is not caused by a temporary loss of generating capacity (e.g. blackout of one generator).

In Fig. 11.1.11 is presented an example of the Single Line Diagram (SLD) for the main power distribution system of a vessel with a Power Management System (PMS). The power plant consists of two 6.6 kV generators connected to MSB No 1, two connected to MSB No 2 and one that can supply either MSB No 1 or 2. The 6.6 kV main switchboards supply cargo switchboards No. 1

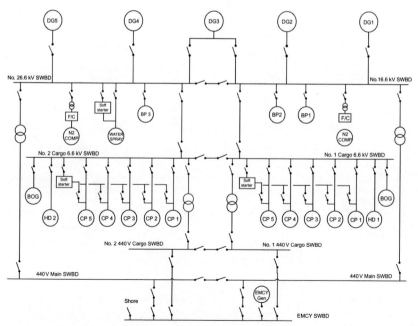

FIG. 11.1.11 Main power distribution system single line diagram. *(Drawing based on Kongsberg documentation.)*

and No. 2, and are also supplying the 400 V switchboard via 6.6 kV/400 V transformers. The 400 V emergency generator is connected to the 400 V main switchboard. The cargo switchboards are supplying 6.6 kV cargo pumps.

11.2 Wartsila Integrated Automation System (WIAS)

11.2.1 General

The Wartsila Integrated Automation system (WIAS) consists of Personal Computers (PCs) with their Monitors for visualisation and control of the processes and Programmable Logic Controllers (PLCs) used as processing units. The computers and controllers are connected via a redundant ring network. A typical WIAS topology is presented in Fig. 11.2.1.

Presented in the Fig. 11.2.1 system topology are two (2) Personal Computers (PCs) named IOServer1 and IOServer2 connected to the redundant process network. The PCs are arranged to work in a redundant manner and simultaneously possess all data, i.e. alarms, trends etc. The system presented in this figure system can communicate with the vessel administrative system, fire detection system and tank gauging system. The Personal Computers work as Operator Stations (OS) and have built in Supervisory Control And Data Acquisition

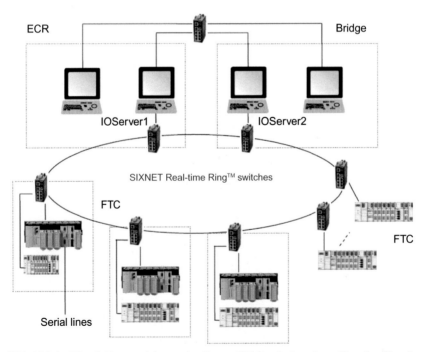

FIG. 11.2.1 Wartsila Integrated Automation System WIAS typical topology. *(Based on Wartsila WIAS General Functional Description.)*

(SCADA) software used for HMI interfacing with the controlled processes. Field Termination Cabinets (FTC) contain Programmable Logic Controllers (PLC) with their Central Processing Units (CPU) and appropriate software.

When the Integrated Automation System WIAS is certified for Unattended Machinery Space (UMS) according to Lloyds Register (LR) rules, or E0 for periodically unattended machinery space according to Det Norske Veritas Germanischer Lloyd (DnVGL) rules then the system drawing in Fig. 11.2.1 is enhanced with Extension Alarm System (EAS) panels installed in Engineer's cabins, Mess, lounge, and Bridge etc. The EAS system is activated on the Operator Station installed in the Engine Control Room (ECR). Additionally the ECR Operator Station allows selection of the Officer on Duty.

The WIAS typical topology can be completed with external alarm rotating lights and horns in machinery areas, activated in case of alarm condition at the Operator Station in ECR.

The Integrated Automation System (IAS) may be completed with a 'Dead Man system' initiating an alarm if the system operator does not acknowledge the Dead Man' request within a predefined time.

Field Termination Cabinets (FTC) are equipped with distributed I/O. Redundant I/O Servers are connected via a redundant network, as presented in Fig. 11.2.2.

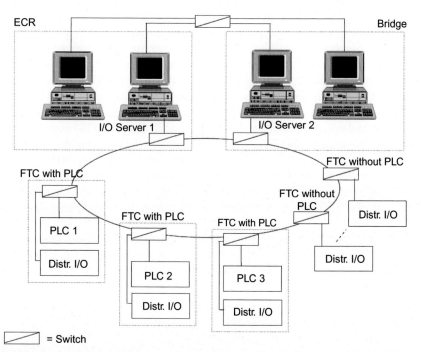

FIG. 11.2.2 Redundant I/O servers and redundant communication. *(Based on Wartsila WIAS General Functional Description.)*

The redundant network presented in Fig. 11.2.2 is made up of a 100 MB Ethernet ring network and connects all PLCs and Server Operation Stations, I/O Server 1 and 2. The switches connected in a ring network topology are working in redundancy, and in normal operation one link between each of the switches is in back-up mode. The back-up link is activated when an error is detected somewhere else on the network. This activation is done by switches without interaction from I/O Server or PLC. Each node (i.e. I/O Server, PLC Cabinet, distributed I/O) is connected on the process network to a separate switch to prevent loss of communication with more than one node if a switch fails. Network switches are supplied from Uninterrupted Power Supply (UPS). Each network switch has its own IP address. The switch contains data with information status monitored by the I/O Server Operator Station.

Client Operator Stations are connected to the Server Operation Stations by a single LAN, 100MB Ethernet network. On the system presented in Fig. 11.2.2 there is installed one I/O Server Operator Station and Client Operator Station in the Engine Control Room (ECR). Secondary I/O Server Operator Station and Client Operator Stations are installed on the navigating bridge.

In such an IAS configuration a single fault does not disable process control either from the Engine Control Room (ECR) or from the bridge. Both I/O Server Operator Stations are redundant. They are able to communicate with each PLC but one I/O Server is configured as the primary while the second is the standby I/O Server. The primary I/O Server handles communication with PLCs and the standby I/O Server is cyclically checking the connections only. The Client Operator Stations and Standby I/O Server Operator Station receive data from the Primary I/O Server. Both I/O Server Operator Stations operate as independent alarm servers displaying and logging alarms and alarm acknowledgement is transferred to other Operator Station as well. The Programmable Logic Controllers (PLCs) communicate with each other independently of the I/O Server Operator Stations and may communicate with more than one distributed I/O node.

The IAS is controlled from the Operator Stations using a graphical interface Supervisory Control and Data Acquisition (SCADA) system. The MIMICS presented on the Operator Station monitors contain overview and sublevel pictures enabling operator control and surveillance of the processes. The pictures contain information on equipment status on/off, alarms, measured parameters and process trends.

Each Field Termination Cabinet (FTC) consists of two remote I/O islands, network switches, PLC and redundant power supply installed in a metal cabinet. Generally, different processes are controlled by different PLCs. Such a distributed architecture minimises loss of process control in case of Programmable Logic Controller (PLC) failure. Additionally, each PLC has self-diagnostic functions.

Analogue input I/O modules support signals of type$\pm$10V,$\pm$5V,0-10V, 1-5V, 4-20mA, -20mA and thermocouple sensors PT100, PT1000, NI1000. Analogue output I/O modules support signals of type $\pm$10V, $0-20$mA and $4-20$mA. Digital I/O modules support input and output signals of 245V DC, 115V AC and 230V AC.

An example of a Wartsila Integrated Automation System (WIAS) Block Diagram is presented in Fig. 11.2.3.

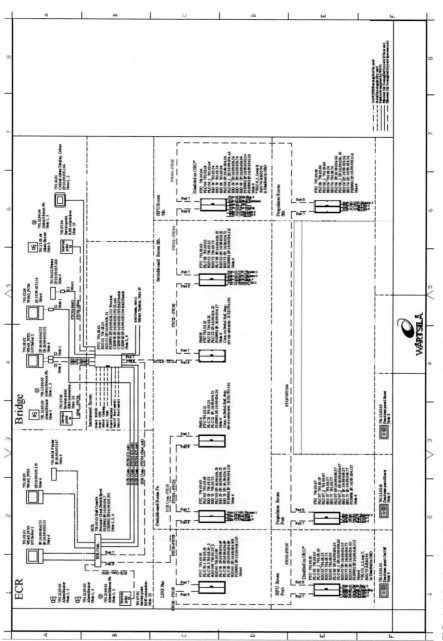

FIG. 11.2.3 Example of WIAS system block diagram. (*Based on Wartsila documentation.*)

The system shown on this WIAS Block Diagram includes WIAS PC01 and PC02 I/O Server Operator Panels and WIAS PC03 and PC04 Client Operator Panels.

The I/O Server Operator Panels are connected in a ring network with: PMS A and a Field Terminal Cabinet (FTC) installed in Switchboard Room Portside, PMS B and a FTC installed in Switchboard Room Starboard, and other FTCs installed in LNG Central, Hydraulic Power Unit (HPU) Room Portside, Propulsion Room Portside, Propulsion Room Starboard, and HPU Room Starboard.

To the Engine Control Room (ECR) Console are connected two (2) Alarm Buzzers, Alarm Silence Pushbutton, a printer and a system Service Point. To the FTC installed in the Instrument Room are connected a Colour Critical Alarm Display (installed on the Bridge), Alarm Buzzer, Alarm Silence pushbutton and system Service Point.

11.2.2 Basic functions of Wartsila Integrated Automation System (WIAS)

The Wartsila Integrated Automation System (WIAS) provides Alarm functions and Control functions. Handling of alarms is presented in Fig. 11.2.4.

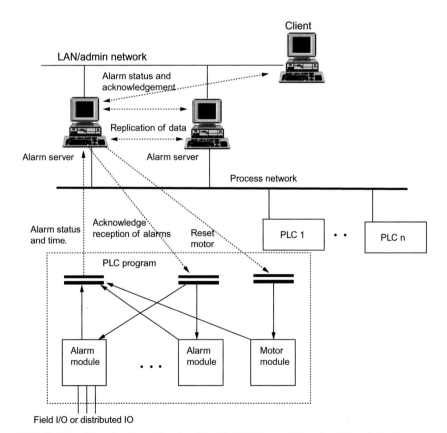

FIG. 11.2.4 Alarm handling. *(Based on Wartsila WIAS General Functional Description.)*

Alarms are in Field I/O or distributed I/O and are transferred to Programmable Logic Controllers (PLCs). From each PLC alarms are transferred to Alarm Server Operator Stations and from there are distributed to Client Operator Stations. Alarms can be acknowledged in the Alarm Server Operator Stations. Alarms are categorised by priority i.e. Critical Alarm, Alarm, Pre-warning, Blocked Alarm and are displayed on Operator Station Monitor(s) as shown in Fig. 11.2.5.

Type of alarm	Colour	
Critical alarm	Magenta	
Alarm	Red	
Pre-warning	Yellow	
Blocked alarm	Green	

FIG. 11.2.5   Alarm list colours. *(Based on Wartsila WIAS General Functional Description.)*

New alarms are displayed on top of the alarm list. Unacknowledged alarms are displayed flashing, and acknowledged alarms are displayed with a firm colour. An alarm remains on the alarm list until the alarm situation return to normal and the alarm is acknowledged. An example of an Alarm page print screen is presented in Fig. 11.2.6.

FIG. 11.2.6   Example of Alarm page print screen. *(Based on Wartsila WIAS General Functional Description.)*

Control functions are carried out from Operator Stations. The control system is divided into several process areas. Each process can be controlled form one Operator Station at a time. It is possible to change the controlling Operator Station by using commands: "Ask for permission to take control", "Take Control" and "Give control". Processes and their elements i.e. motors, pumps, fans, valves etc. are controlled from the Operator Station in command or automatically by PLC. The Operator Station receives feedback signals from the switchboards to indicate the status of process elements i.e. running, speed, fault, not available etc. Pumps are configured as Duty or Standby.

A substantial part of the control system is the Power Management System (PMS). The main functions of a power management system are: connecting and disconnecting of diesel generators depending on required load, automatic generator changeover when one of the generators is tripped, preventing generator overload by thrusters and main propulsion converters, connecting heavy consumers, load shedding, blackout prevention and manual operation of breakers and generators. An example of an Integrated Automation System (IAS) including power management functions is shown in Fig. 11.2.7.

The Power Management System (PMS) presented in Fig. 11.2.7 consists of two (2) dedicated Programmable Logic Controllers (PLCs) connected in the Integrated Automation System (IAS) ring. Both PLCs are installed in Main Switchboards (MSBs): PMS A is installed in MSB Port Side and PMS B is installed in MSB Starboard. When a fault occurred on PMS A then PMS B takes control of all Field Terminal Cabinets (FTC) controlled by PMS A. All FTCs are connected in a ring configuration and are collecting control data from hardwired I/O's.

In the Power Management System (PMS) the following generator data is measured: Active power P (kW), Reactive power Q (kVAr), Voltage U (V), Current I (A), frequency f (Hz) and Power Factor (cosφ). These values are presented on MIMIC diagrams on the Operator Station monitors.

An example of a PMS System Block Diagram is presented in Fig. 11.2.8.

11.3 Wartsila Integrated Automation System (IAS) NACOS Valmatic Platinum

The Wartsila Integrated Automation System (IAS) for Navigation, Automation, Control NACOS VALMATIC Platinum has modular architecture that enables development of redundant and flexible control and monitoring systems for ships and offshore units. The NACOS system can be used for: Machinery control and monitoring, Power Management, Propulsion & Thruster control, Heating, Ventilation and Air Conditioning (HVAC), Emergency Shut Down (ESD), control and monitoring of ships cargo including Liquefied Natural Gas (LNG), Navigation, auxiliary equipment control and monitoring and Dynamic Positioning.

The NACOS VALMATIC Platinum Integrated Automation System presented in Fig. 11.3.1 consists of two (2) Operator Stations (OP) connected by a redundant fibre optic ring network to the Information Management System

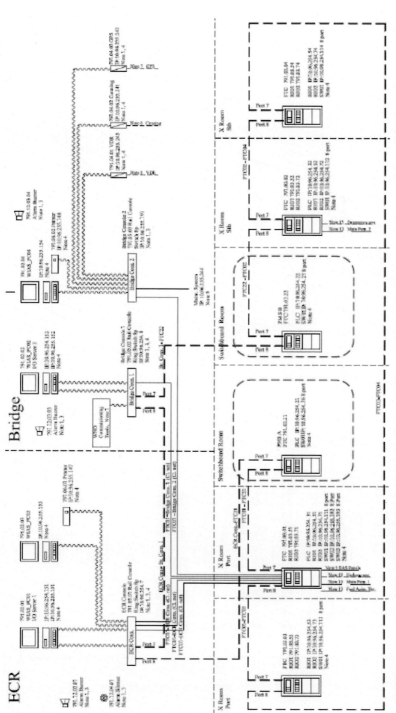

FIG. 11.2.7   Example of Integrated Automation System (IAS) with Power Management System (PMS). *(Based on Wartsila WIAS General Functional Description.)*

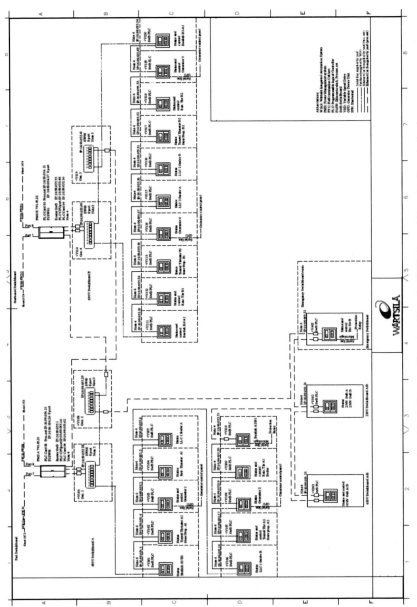

FIG. 11.2.8 Example of PMS system block diagram. (*Based on Wartsila WIAS General Function Description.*)

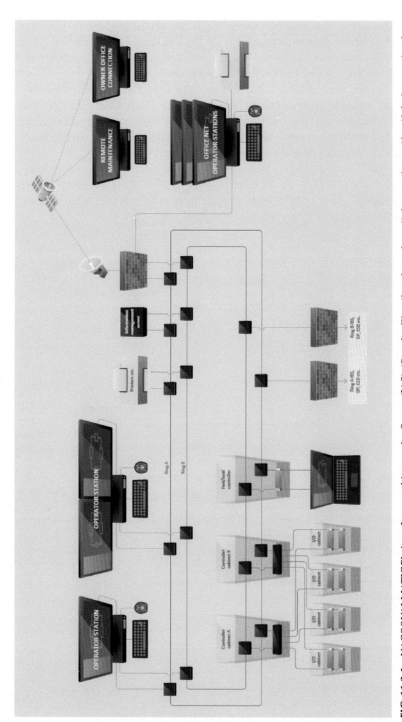

FIG. 11.3.1 NACOS VALMATIC Platinum Integrated Automation System (IAS). (*Based on Wartsila webpage https://cdn.wartsila.com/docs/default-source/product-files/ea/automation/brochure-o-ea-valmatic.pdf?um_source=autnavdp&utm_medium=automation&utm_term=valmaticplatinum&utm_content=brochure&utm_campaign=msleadscoring.*)

(IMS), Redundant Controller Cabinets 'A', 'B' and the Redundant I/O Cabinets connected to them, plus redundant rings for DP, ESD and similar functions. The system may be connected via the internet with a Remote Maintenance Server, Office Operator Stations and Owner Office Server.

The system main attributes include the following functions: Redundancy with dual redundant ring networks, Programmable Logic Controllers (PLCs) and I/O's, local control and operation at processor levels, hot swap of any computers, fast system recovery, remote diagnostic functions, data logging functions with trend analysis and online/offline commissioning.

The NACOS VALMATIC Platinum system integrates the control of main machinery systems such as: Propulsion plant, Power plant including Power Management System (PMS), Fuel plant with bunker stations, Bilge system, Ballast system and Auxiliary machinery control.

Chapter 12

Safety management systems

Chapter outline

Abstract

This chapter describes safety management systems installed onboard ships and offshore units. The described Fire & Gas Detection system include examples of Fire and Gas Philosophy and Cause & Effect Diagrams. Also described is an Emergency Shutdown System supplemented by an Emergency Shutdown Philosophy. Additionally in this chapter is presented a Public Address/General Alarm System together with its typical block diagram.

12.1 General

The Safety Management Systems should present in a clear and structured way information regarding vessel critical situations to allow the crew to take the proper decisions and actions. Safety systems typically include Fire & Gas Detection (F&G), Emergency Shut Down (ESD), Watertight doors and CCTV. Often Safety Integrated Level (SIL) is determined for Safety Management subsystems and on owners or maritime administration request they may be certified for SIL levels—see Chapter 7.

12.2 Fire & gas detection (F&G)

An approved Fire and Gas (F&G) detection system is required by the vessel flag administration and the classification society that issues class and statutory certificates. Technical requirements for offshore units are included in IMO MODU

Ship and Mobile Offshore Unit Automation. https://doi.org/10.1016/B978-0-12-818723-4.00012-0

Code, while those for merchant vessels and offshore support vessels are found in the IMO SOLAS Convention. Additional requirements to those imposed by IMO and the flag administration are found in the Classification Societies rules. In addition, specific customers (e.g. oil companies) and coastal states (e.g. Norwegian Maritime Administration) may have their own special standards or requirements.

The F&G system consists of: fire, smoke (and sometimes gas) detection equipment and manually activated call points installed in all accommodation spaces and in machinery spaces where required by vessel administration. On offshore vessels, smoke detectors are sometimes required at HVAC inlets. Similarly, gas detectors are often fitted at HVAC inlets, and are required to continuously monitor enclosed areas in which an accumulation of flammable gas or Hydrogen Sulphide (H2S) gas may occur. If the offshore vessel includes hydrocarbon production or drilling equipment, then gas detection equipment must be provided in these areas.

The F&G system provides early fire and gas detection and initiates protective actions automatically or manually by crew. Often the F&G systems are arranged to automatically release firefighting systems and control emergency ventilation systems.

F&G detection system inputs (e.g. detectors for Smoke, Heat, Gas, Flame and Manual Call Points) are used to initiate F&G system outputs according to predesigned Cause & Effect (C&E) diagrams. To avoid any spurious actions or shutdown situations in critical areas the F&G system outputs are often based on a '2oo2 or 2oo3' voting principle applied to the inputs.

Fire & Gas systems are developed in multiple redundant configurations: utilising redundant computers, I/O, net and power supplies. The systems are built with different Safety Integrity Levels (SIL) i.e. SIL1, or SIL2, or SIL3. F&G systems are provided with self-monitoring facility including monitoring of lines/loops and individual detectors.

The Fire and Gas detection logic is usually outlined in a F&G Philosophy document, which is approved by the customer and classification society in advance of the system engineering process. Approval of the Cause & Effect Matrix is also required. Late changes to the philosophy and Cause & Effect logic can have expensive consequences for a project.

12.2.1 F&G system philosophy

The Operations Manual for a Vessel or Offshore unit should also contain the F&G System Philosophy. The philosophy will list all applicable Regulations, Codes, Rules and Standards and will describe the F&G system interfaces with the Emergency Shut Down system (ESD). In the philosophy will be listed the main objectives of the Fire and Gas Detection System (F&G); e.g. early and reliable detection of fire and gas hazards, activate early warning, alert personnel and initiate protective actions manually or automatically.

The philosophy for system voting on F&G system outputs to avoid any spurious shutdown situations (e.g. '2oo2 or 2oo3' voting principle) should be clearly described.

The F&G System Philosophy should also describe in detail the requirements and principles for: system and component redundancy, system self-checking and diagnostic capabilities. No single failure should impair the F&G system from detecting and activating the safety functions.

The F&G System Philosophy will describe communication with Visual Display Units (VDU) at the Operator Stations in Engine Control Room (ECR) and on the Bridge.

The F&G system mimic screens will display information on detector or line fault, fire alarm, detector inhibition and suppression status.

The scope and form of a typical fire & gas system philosophy is given below[1]:

---

1. **Purpose**
   The purpose of this philosophy is to outline the basis for Fire and Gas Detection System, Shutdown System and their interfaces.
2. **Objective**
   Fire and Gas System and Shutdown System shall ensure the safety of the personnel, vessel and equipment. Both systems shall minimise all consequences in emergency in case of fire or gas.
3. **Regulation, Codes and Standards**
   The following regulations are applicable to the vessel:
   - *Lloyds Rules and Regulations for the Classification of Mobile Offshore Units, Part 7, Safety Systems, Hazardous Area and Fire;*
   - *Norwegian Petroleum Directorate (NPD) regulation 'Safety and Communication Systems on Installation in the Petroleum Activities';*
   - *Norwegian Maritime Directorate (NMD) regulation for mobile offshore units.*
4. **Fire and Gas Detection System**
   4.1 General
   The main objectives of F&G Detection System are to provide early and reliable fire and gas hazards detection, activate personnel early warning and alert so protective actions are carried out manually by operators or automatically.
   F&G Detection System shall initiate fire-fighting system, activate alarms on the vessel and shall interface to shut down signals of IAS.
   All Fire Areas on the vessel with potential fire risk are to be provided with appropriate number of Fire Detectors.
   All areas where gas may cause a hazardous situation are to be provided with appropriate number of Gas Detectors.
   F&G Detection system is linked with HVAC system to prevent escalation of fire by isolating effected by fire or gas areas.
   4.2 System description
   F&G Detection system is a redundant and fail-safe system with self-checking and diagnostic capabilities. Single failure shall not impair the system from detecting and activating safety functions.
   Smoke, Heat, Flame and Manual Call Points (MCP) are connected to addressable fire loops and Gas Detectors are connected directly to Fire Alarm Panel.

---

1. Based on Kongsberg project

F&G Detection System initiates fans and dampers shutdown and provides hardwired signals to appropriate equipment shut down. Fire pumps are started automatically on loss of pressure in the system. This activation is based on 2oo2 voting principle and for Diesel Generator Rooms ventilating air inlet ducts is based on 2oo3 voting principle.

F&G Detection System displays its status in Engine Control Room (ECR) and on bridge on Visual Display Units (VDU) and MIMIC Panels.

F&G Detection System displays graphic presentation of fire areas, detectors numbers with location, type of detected hazard and system status including detector fault, inhibit and isolation. Additionally, system records all alarms, detector inhibits and isolations and system failures.

Engine Control Room (ECR) shall be the firefighting main station and shall have all F&G facilities and HVAC control panel. ECR ventilation system air intake ducts shall be equipped with Gas Detectors to totally isolate ECR upon gas detection.

4.3 System overview

The following Fire Areas are identified on the vessel:

1. Diesel Engine Room 1—with MCP and Smoke, Flame, HC Gas Detectors
2. Diesel Engine Room II—with MCP and Smoke, Flame, HC Gas Detectors
3. Engine Control Room—with MCP and Smoke, HC Gas Detectors
4. Engine Utility Room—with MCP and Smoke, Heat, Flame, HC Gas Detectors
5. Air Compressor Room—with MCP and Smoke, HC Gas Detectors
6. High and Low Voltage Electrical Room I
7. High and Low Voltage Electrical Room II
8. Emergency Generator Room—with MCP and Smoke, HC Gas Detectors
9. Accommodation Fan Rooms—with MCP and Smoke, HC Gas Detectors
10. Accommodation Area I—with MCP and Smoke Detectors
11. Accommodation Area II—with MCP and Smoke Detectors
12. Accommodation Area III—with MCP and Smoke Detectors
13. Accommodation Area VI—with MCP and Smoke Detectors
14. Accommodation Area V—with MCP and Smoke Detectors
15. Accommodation Area VI—with MCP and Smoke Detectors
16. Galley—with MCP and Heat Detectors
17. Laundry—with MCP and Smoke Detectors

4.4 Cause & Effect Diagrams

For all Fire Areas a Cause & Effect Diagram shall be developed and attached to Fire & Gas System Philosophy—see next Chapter.

5. **Fire water Pumps**

Fire Water Pumps shall start automatically on low pressure in main manifold.

6. **Shutdown System**

Shut down system minimises the consequences of detected by Fire & Gas Detection System abnormal condition. The system shall be fail safe, self-checking and self-diagnosing to indicate any system failures. Single failure shall not impair the system from activating safety functions.

Shut down functions are initiated by Fire & Gas system or manually and isolates affected ventilation areas and certain Non-Ex equipment, initiates General Alarm (GA) and Public Address (PA), shutdown Boilers and manually shutdown cranes.

12.2.2 Cause and effect diagrams

C&E diagrams are developed for each separate identified Fire Area. The boundary of a Fire Area is defined on the basis of fire protection systems, HVAC areas and physical layout of boundaries.

An example of a typical Cause & Effect diagram is shown in Fig. 12.2.1.

12.3 Emergency shut down system (ESD)

An automatic Emergency Shut Down System (ESD) is installed on most offshore units to prevent disastrous fires and explosions and minimise their consequences. The emergency shutdown system interconnects vessel or offshore unit controls in such a way that the necessary installations are automatically isolated and shut down to a safe level of operation either through manual initiation by the operator or by safety system instrumentation.

Emergency Shut Down systems minimise the consequences of the emergency situations caused by presence of hydrocarbons or fire in nominally safe areas. Typically, an activated Emergency Shut Down system shuts down ventilation systems and fuel supply, and isolates electrical equipment to prevent escalation of hazard events.

An ESD Philosophy document should be developed at an early stage of a project. Various regulatory bodies have specific requirements for the different ESD levels and associated action, which may sometimes conflict with the desires of customers. In is important that a unified approach is agreed before detailed engineering is started. This document will describe the ESD system interfaces with F&G system, IAS system and the PA&GA system.

When this document is finalised, care must be taken to ensure that all equipment which is to remain live in a potentially hazardous gas atmosphere after a shutdown is rated appropriately for the explosion risk.

The Emergency Shut Down System has interfaces to: fire and gas detection system, alarm and monitoring system, PA/GA communication system and HVAC.

The system should be designed so that a single fault cannot prevent (or cause) a shutdown initiation and the ESD system should have loop monitoring functions. The Emergency Shut Down system power shall be supplied from two sources: emergency power system and from uninterruptible power supply (UPS). It should be possible to test the ESD system without interrupting other vessel systems during vessel operation. When the ESD system is activated, automatic attempts to re-start machinery are disabled.

After shutting down, the following systems should normally remain operable; emergency lighting (for half an hour or more depending on agreed philosophy or rules), Public Address and General Alarm PAGA, and distress and safety radio communications.

Manual activation of the ESD system should be possible from at least two locations which are outside hazardous areas, and also at the backup control station of a DP3 offshore unit.

		Effects								
	F&G actions						Vessel action			
Alarm location	General Alarm / PA	HVAC	Emerg. Vent Start	Doors	Isolate Non Ex	VMS	Close / lift, gangway	Muster	Move off	
Yes	Auto or not acnowledged within 2 min	Trip and interlock galley, lower deck, laundry and accomodation HVAC	Yes	Yes	No	Fire and gas alarm	On command	On command	On command	
Yes	Not acknowledged within 1 min	No	No	No	No	Fire and gas alarm	No	No	No	
Yes	Auto or not acnowledged within 2 min	Trip and interlock galley, lower deck, laundry and accomodation HVAC	Yes	Yes	No	Fire and gas alarm	On command	On command	On command	
Yes	No	No	No	No	Yes	Fire and gas alarm	No	No	No	
Yes	Yes	Trip and interlock galley, lower deck, laundry and accomodation HVAC	No	Yes	Yes	Fire and gas alarm Stop air Compressors	On command	On command	On command	
If fail open or close	No	Interlock on fauled HVAC	No	No	No	No	No	No	No	

FIG. 12.2.1 Cause & Effect diagram for Area No 1 Accommodation Fan Room.

Abandon vessel/platform shutdown (APS) push-buttons are usually located in main and emergency control stations e.g. bridge and back-up DP control room, muster stations, lifeboat stations and helicopter deck.

Additional local Emergency Shut Down pushbuttons are usually installed on vessel/offshore units in the Engine Room to shut down main engines, fuel oil pumps, oil separators, oil transfer pumps related fans and to close ventilation dampers related to Engine Room area. On Helideck, there is usually installed a local ESD pushbutton to stop the helifuel transfer pump.

12.3.1   Emergency shut down philosophy

The Emergency Shut Down Philosophy outlines the design basics of electrical equipment which shall be shut down or maintained live to minimise consequences of unexpected release of hydrocarbons and associated risk of fire or explosion. An ESD philosophy defines different shutdown levels e.g. ESD Level 1, ESD Level 2 and APS (Abandon Platform Shutdown) Level. The exact naming and definition of these levels may vary from project to project.

An ESD Cause and Effect Diagram or Matrix is normally included in the ESD philosophy document at the start of project engineering.

12.4   Public address/general alarm systems (PA/GA)

The Public Address and General Alarm system provides clearly audible announcements of alarm situations and provides an internal means of communication to transfer information between all spaces where action may be necessary in case of an emergency.

The PA/GA system distributes audio alarm, visual alarms and emergency messages. Audible alarms are supplemented with visual display units e.g. rotating lights.

The Public Address/General Alarm system is interfaced with other systems such as:

– Fire & Gas (F&G) system for alarm input;
– UHF interface for transferring emergency messages;
– PABX for transferring phone information;
– Entertainment interface for muting TV and other sources when Emergency messages and alarms are announced.

The PA/GA system generates different signals, typically for following different situations:

– General alarm (GA);
– Prepare to abandon platform vessel/offshore unit alarm (PAPA);
– Toxic gas detected.

The system is normally divided for different zones, e.g. Crew accommodation, guest accommodation, public areas, Engine Room areas and deck areas including lifeboat muster areas. The PA/GA system is usually fully redundant in design and installation including double amplifiers, speakers and networks (A and B cable routes etc.).

An example PA/GA Block Diagram is shown in Fig. 12.4.1.

An example of PA/GA System Architecture is shown in Fig. 12.4.2.

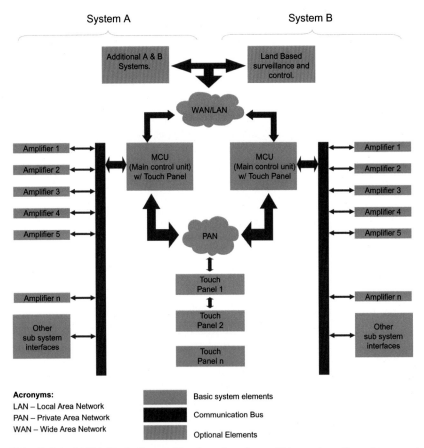

FIG. 12.4.1 PA/GA block diagram.*(Brochure from Jotron Webpage https://www.jotron.com/ Artikkel/Brochures/Maritime-&-Energy/10002032.php.)*

Safety management systems **Chapter** | 12 **131**

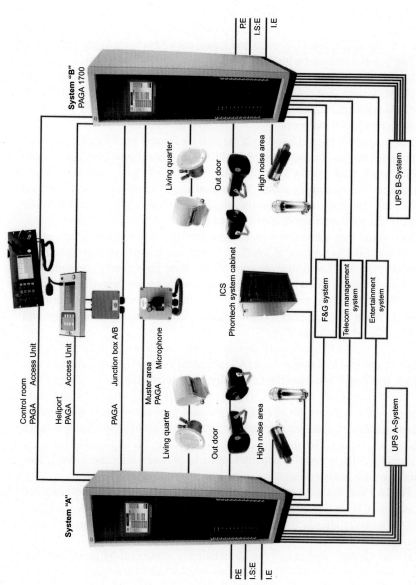

FIG. 12.4.2 PA/GA system architecture. (*Based on Jotron Marine & Energy Division Product Catalogue http://pdf.nauticexpo. com/pdf/jotron/product-catalogue-july-2017/22206-100695.html.*)

12.5 Interface with HVAC system control

Required Heating, Ventilation and Air Conditioning (HVAC) functions are described in the vessel or offshore unit Technical Specification or a separate HVAC Philosophy document. The HVAC system is controlled to some extent by the IAS, depending on the degree of control provided by HVAC vendor, and is linked with the Fire & Gas (F&G) System and Emergency Shutdown System (ESD).

The IAS is connected with the HVAC Motor Control Centers (MCC) to control and monitor fans, perhaps via a dedicated HVAC control panel supplied by the HVAC vendor. Fan status information is sent to F&G System. On confirmed Fire & Gas Alarm, the ESD system will shut down applicable fans and open/close applicable fire dampers.

Single speed fans are linked with the IAS for remote Start/Stop commands, running indication, common fault, Local/Remote control selection, and current measurement. Fans can be controlled start/stop locally and stopped in emergency from the machinery room. When dual single speed fans are controlled by IAS than the duty fan is normally running. When the duty fan fails then the stand by fan shall start automatically and an alarm signal is generated.

Two speed fans are linked with the IAS for Start/Stop—low and high speed commands, and indication of low/high speed mode and running indication, common fault and Local/Remote control selection. Fans can be controlled for start/stop, low/high speed locally and stopped in emergency from the machinery room. When dual fans are controlled by IAS than the fan on duty is normally running. When duty fan fails then the standby fan shall start automatically and an alarm signal is generated.

Variable speed fans are controlled by a Variable Frequency Drive (VFD) locally. Start/stop signal is sent from IAS. Speed signal and common alarm is sent to IAS. It is possible to stop the fan locally in an emergency.

The IAS provides interlocks so that specific fans are stopped when main inlet or outlet dampers are closed, and exhaust fans are stopped when intake fans in machinery area are stopped, or intake fans are stopped when exhaust fans are stopped, and when main ventilation fails than backup ventilation starts (if provided). There are also usually interlocks between inlet and exhaust fans in the accommodation area.

The IAS will initiate the start/stop of Air Handling Units (AHU). Automated control of AHU is carried out by their Local Control Panels (LCP). Typically, the IAS receives signals for AHU(s) common alarm, frost protection and fun running.

HVAC chiller compressors are automatically controlled by pressure transducers connected to their LCP and they are interlocked with the air conditioning units. The compressors are automatically stopped if there is no cooling water flow. Chiller water pumps are started by IAS and running and fault signals are sent to IAS.

Chapter 13

Position keeping systems

Chapter outline

Abstract

In this chapter are outlined position keeping requirements and two types of vessel and offshore unit position keeping systems are described. Firstly—Dynamic Positioning for vessel or offshore units with thruster assistance and Secondly—Position Mooring Systems using anchors to keep position. An example of machinery automation system integrated with Dynamic Positioning system is presented.

13.1 General

Most offshore vessels have a high requirement to maintain relative or absolute position and/or heading. In the simplest cases, this is provided by means of a multiple line spread mooring system. In some cases, these moorings are supported by additional power from thruster assist systems. Finally, full dynamic positioning (DP) systems are available to permit operations independently of any mooring equipment. DP is particularly important for operations in deeper water where mooring is difficult.

In the case of DP vessels, the position and heading are assured by a computer based Dynamic Positioning system controlling the vessel's propellers and thrusters. Different degrees of redundancy may be built into these systems in order to achieve an acceptable level of risk with regard to loss of position. The degree of redundancy and reliability required depends on the nature of the operation being conducted. Operations involving human beings and high pressure hydrocarbons (e.g. saturation diving or drilling) and operations nearby to fixed installations demand a higher level of reliability than those conducted in open water with no personnel or hydrocarbon risk (e.g. Remotely Operated Vehicle—ROV survey of a pipeline).

Ship and Mobile Offshore Unit Automation. https://doi.org/10.1016/B978-0-12-818723-4.00013-2

Different levels of position keeping reliability are defined by IMO MSC Circular 645 'Guidelines for Vessel's with Dynamic Positioning systems' as follows;

- Class 1—where loss of position may occur in the event of single failure
- Class 2—where loss of position is not to occur in the event of a single fault in any active component
- Class 3—where loss of position is not to occur in the event of a single fault in any active component or system, or loss of any compartment

To meet Class 3 standard, vessels must be designed and built with DP, power and propulsion systems separated by fire and watertight boundaries so that loss of any one space (e.g. an engine room) will permit the remaining systems to continue to maintain vessel position. Such vessels have multiple engine rooms and switchboard rooms, and separated cable routes.

To meet Class 2 standard, it is only necessary that the systems may continue to function and position will be maintained in case of loss of a single component such as a generator, a DP controller or a thruster. Multiple engines, thrusters and controllers are therefore installed, but there may be only one engine room or one switchboard room.

Each classification society has developed its own rules for their class notations, which follow the principles of the IMO criteria. The relationship between these different notations is shown in Fig. 13.1.1.

	DnVGL	ABS	LR
IMO Class 1	DYNPOS-AUT	DPS-1	A
IMO Class 2	DYNPOS-AUTR	DPS-2	AA
IMO Class 3	DYNPOS-AUTRO	DPS-3	AAA

FIG. 13.1.1 DnVGL, ABS, LRS class notations.

13.2 Dynamic positioning system

Unless it is moored, a vessel or offshore unit may maintain its position and heading using its own thrusters and propellers controlled by the Dynamic Positioning System (DP).

Thruster and propeller actions are controlled by the DP system which receives position/motion data from a range of input sensors. These can include absolute position indicating systems such as satellite navigation (e.g. DGPS), and other methods such as taut wire (in relatively shallow water), hydro acoustic transponders (e.g. HIPAP), and other relative position indicating systems using laser or radar reflection, which require the presence of a nearby target platform.

Signals from these sensors are corrected for the effect of vessel motion by reference systems measuring 'Surge', 'Sway', 'Pitch', 'Roll' and 'Heave', together with yaw input from a gyro compass system.

The DP system has different operational modes including Auto Heading mode (e.g. for a drillship in open water), Auto Position mode, Joystick mode, Follow Target mode or Auto Track mode (e.g. for pipelaying).

The ship/offshore unit generates thrust to maintain position and/or heading by means of its main propulsion system (usually fitted with controllable pitch propellers if direct driven by a diesel engine), high lift rudders working in conjunction with the propellers, and independent thrusters with related drive units and thruster controllers. Thrusters may be of the azimuthing type, or simple transverse thrusters (usually fitted in tunnels). The various propulsion devices have their azimuth heading (if not fixed) and output thrust (adjusted by varying propeller speed or pitch) controlled by means of commands generated within the dynamic positioning system. Capability to adjust propulsion or rotational speed may be provided by a variable speed electric motor, or a direct coupled diesel engine.

The DP system is closely interlinked with the Propulsion/Steering/Thruster systems and the power system. The power system includes the prime movers with their auxiliary systems, generators, switchboards and distributing system. The overall integrated system is designed to keep the vessel or offshore unit within specified position and heading limits, up to the weather limits of the power generation and propulsion systems. The Power Management System will work actively during such operations to ensure that available power generation matches the fluctuating demands from the propulsion system.

The DP control system consists of a computer system containing the vessel/offshore unit mathematical model, the various sensor systems, operator panels and position reference systems.

An example of the DP and Thruster Control systems forming part of an Integrated Automation System (IAS) is shown in Fig. 13.2.1.

An example of a Thruster Control system is shown in Fig. 13.2.2.

Often the Machinery Automation system is integrated with the Dynamic Positioning system as is shown in Fig. 13.2.3.

The integrated system example presented in Fig. 13.2.3 consists of six Machinery Automation Operator Stations (OS), four Dynamic Positioning Operator Stations, two Thruster Operation Stations, one Joystick controller, one Fire Alarm Operation Station and two Hydro Acoustic Position Reference Operation Stations, Helideck Monitoring System, eleven Watch Call Panels and eight Operator Fitness Panels.

The Machinery Automation Operator Stations allow control of auxiliary machinery including: Power Management System (PMS), ballast and bilge systems and auxiliary machinery control. Operator Stations HMI mimics illustrate all alarms and values readings from the alarm and monitoring system.

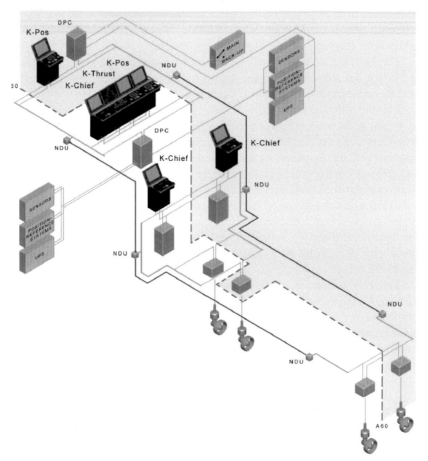

FIG. 13.2.1 Integrated Automation System (IAS) including DP and Thruster Control systems. *(Kongsberg webpage for K-Pos DP Product Description https://www.km.kongsberg.com/ks/web/ nokbg0240.nsf/AllWeb/14E17775E088ADC2C1256A4700319B04?OpenDocument.)*

Power management control inputs can be carried out using the Operator Stations or can be controlled through manual controls on the appropriate switch-boards panels.

The ballast control system (usually part of the IAS) is used for vessel/off-shore unit trimming, ballasting and de-ballasting. The ballast control functions include: ballast tank level and monitoring, trim/list corrections of tank level/vol-ume, ballast tank volume presentation, remote start/stop of electrically driven ballast pumps, remote control of valves, monitoring of hydraulic power pack for valve control, draught measurement and trim/list calculations.

A bilge control system is used for drainage of dry compartments of the ves-sel or offshore unit. Bilge control systems include: bilge tank level monitoring, bilge tanks volume presentation, monitoring of bilge wells, level alarms, remote

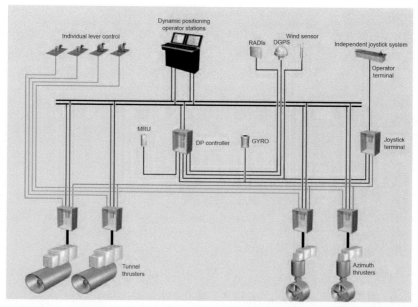

FIG. 13.2.2   Thruster Control system. *(From Kongsberg webpage https://www.km.kongsberg. com/ks/web/nokbg0240.nsf/AllWeb/0A0C3F74B421A7DEC1256A49002DA456?OpenDocument)*

start/stop of air driven ejectors and electrically driven bilge pumps, remote control of valves and monitoring of hydraulic power pack for valve control.

Auxiliary machinery control systems include control of: fuel oil system, sea water cooling system, fresh water cooling system, potable water system, deck drain system, watertight integrity, compressed air system, waste oil system and ventilation systems (HVAC).

Thruster Operator Stations permit the selection of control between lever control, DP control system and independent Joystick system, allow command transfer between alternative control stations, and also respond to Operator manual control inputs via the thruster levers (if selected). Thruster Operator Stations are linked to the steering and autopilot systems for normal—not DP navigation. Manual thruster control or joystick control is sometimes used in harbour manoeuvring situations.

The Dynamic Positioning Operator Stations allow control and monitoring of the vessel/offshore unit position and heading. The operator may select the different modes of DP operation, the different sensors to be input to the system (or deselected) and the setting of alarm limits etc.

The Dynamic Positioning system obtains necessary position data from Global Navigation Satellite System (GNSS) via GPS L1/L2, IALA, SPOTBEAM, GLONASS, UHF and INMARSAT antennae. The system obtains vessel movement and environment data from Main and Back-up DP sensors including Gyro, Motion Reference Units MRU and wind sensors.

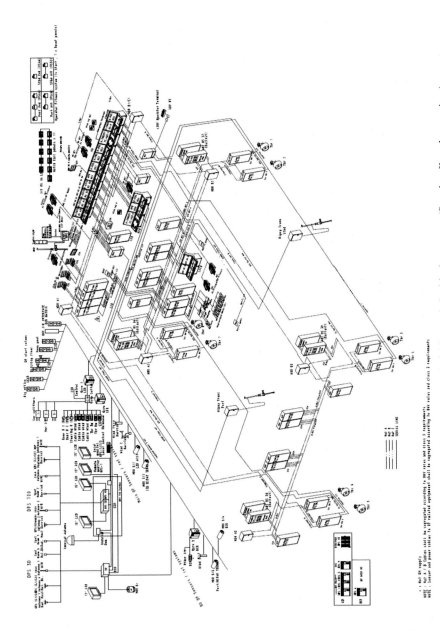

FIG. 13.2.3 Machinery Automation system integrated with Dynamic Positioning system. (*Based on Kongsberg project.*)

The Machinery Automation system integrated with Dynamic Positioning is typically interlinked by:

– Fibre Net A connecting PS Field Stations (FS) to Network Distribution Units (NDU) and Operator Stations (OS)
– Fibre Net B connecting SB Field Stations (FS) to Network Distribution Units (NDU) and Operator Stations (OS)
– Fibre Net C connecting Operator Stations (OS) in CCR to Network Distribution Unit (NDU) and Printers in CCR, LECR
– Serial Bus connecting Thrusters Field Stations (FS) to Operator Stations (OS) in CCR and LECR
– Fibre Net for High Precision Acoustic Positioning (HiPAP) connected to Hydro acoustic Positioning Reference (HPR) system
– Network for connecting Deferential Positioning System to GPS L1/L2, IALA, SPOTBEAM, GLONASS, UHF and INMARSAT antennas
– Network for connecting Helideck Monitoring System with Acoustic Doppler Current Profiler, Temperature, Humidity, Visibility, Clouds high sensors
– Network for Watch Call Units (WCU)
– Network for Operator Fitness Units (OFU)

13.3 Position mooring systems

Some vessels such as drilling rigs are fitted with multiple line spread mooring systems. In some cases, these mooring systems are the only means of station keeping, while in other cases they work in conjunction with thruster systems.

A typical mobile unit position mooring system consists of mooring winches and fairleaders, mooring wires/chains with anchors, together with the necessary power supply and control systems including control cabins.

A high power position mooring winch system may be electrically or hydraulically driven. In case of hydraulic powered winches, typically four electrically driven hydraulic power packs are installed on the vessel—one at each corner—to avoid long runs of hydraulic piping through the vessel.

Earlier generation drilling rigs tended to use DC motors of the same type and 1000hp rating as those used in the drilling systems (mud pumps, rotary table etc) for the mooring winches. Modern electric winch systems are typically supplied from medium voltage 6.6 or 11kV AC switchboards.

Electrically driven winch motors operate typically at 690 V and are supplied from mooring winch power supply transformers via Variable Frequency Drives (VFD). In the VFD AC energy is rectified to DC and then is converted to AC with the required frequency Insulated Gate Bipolar Transistors (IGBT). The VFDs need to be located in well protected and air-conditioned spaces to avoid failures caused by high temperature effects on sensitive components such as PLCs, control and measuring equipment.

In order to handle the potential for reverse power created when the mooring line is being paid out under gravity load, the VFDs are usually connected to a braking resistor that absorbs the VFD DC bus back energy created by the winch motor rotating field in the stator. The resistor disposes of energy by heating large volumes of blown air or smaller volumes of circulated cooling water.

The mooring winches are fitted with automatic over speed protection, line tension measuring equipment, line out measuring system and speed control for mooring line pay-out and retrieval.

A winch control cabin is usually located at each of the four corners of a mobile offshore unit. The winch control cabinet is interfaced with the Dynamic Positioning system (if fitted) and the Emergency Shutdown Systems (in some vessels, emergency release of some mooring lines may be required under some accident scenarios). Indication and control of the mooring winches from a central position (usually the Bridge or CCR) is also usually provided on modern mobile offshore units.

An example of anchor winch supply and control block diagram is provided in Fig. 13.3.1:

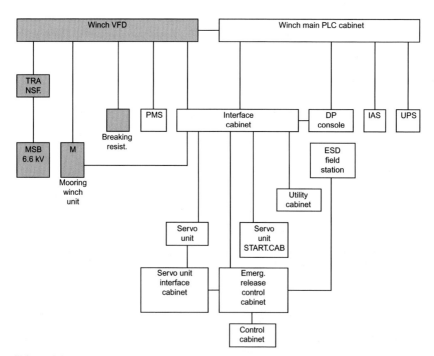

FIG. 13.3.1 Anchor Winch supply and control block diagram.

Chapter 14

Remote diagnostic systems

Chapter outline

Abstract

In this chapter are presented Kongsberg, ABB and Wartsila Remote Diagnostic Systems as installed on many modern vessels and offshore units. The concept of such remote diagnostic systems is described as is the necessity of signing an associated service agreement with service companies employing specialists working onshore and able to help the vessel crew.

14.1 General

Currently, onboard control and monitoring systems are complicated and sophisticated. Crews are not always able to find the reasons for automation system faults. In order to help crews and maintain unit operability the owners are often installing condition monitoring equipment to collect data of control and monitoring system elements aboard and send them for diagnostic services ashore.

For the crew the main advantages of installing remote diagnostic systems are:

- immediate assistance from experts, product engineers and support engineers in trouble shooting and fault finding of faulty control and monitoring system/elements,
- early warnings of detected defects to allow crew to initiate preventive actions,
- Enabling of proactive maintenance activities by detecting potential failure reasons before they escalate and cause automation system failure.

Remote diagnostic systems are profitable for the owners of vessel and offshore units as well; for example, such systems should:

- Maximize vessel uptime, performance and profitability,
- Lower service costs by early warnings of potential failure causes,
- Reduce fault finding time to increase vessel/offshore unit performance.

Ship and Mobile Offshore Unit Automation. https://doi.org/10.1016/B978-0-12-818723-4.00014-4

141

14.2   Kongsberg remote diagnostic system

Kongsberg Information Management System (K-IMS) includes Remote Diagnostic System (RDS) and additional modules such as: Crowd Management system, Ship performance monitoring, Decision support system and Malware protection. Hardware of K-IMS includes portal integrating Kongsberg and third party suppliers applications, data logging equipment secure network, malware protection and web services. By remote access to the control and monitoring system Kongsberg experts analyse equipment and processes trends, alarms and events and are able to format appropriate reports for the crew and the Kongsberg Maritime Global Customer Support Center. Online service engineer analyses received data and provide guidance and support for an operator of onboard systems. Kongsberg K-IMS Remote diagnostic system is presented in Fig. 14.2.1 Kongsberg K-IMS Remote diagnostic system.

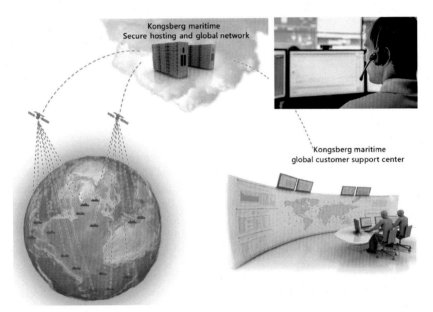

FIG. 14.2.1   Kongsberg K-IMS Remote diagnostic system. *(Based on Kongsberg webpage https://www.km.kongsberg.com/ks/web/nokbg0240.nsf/AllWeb/F5D0D79A0023B87CC12579EA00 43A744?OpenDocument.)*

14.3   ABB Ability™ marine remote diagnostic system

The ABB Ability™ Marine Remote Diagnostic System (RDS) is a diagnostic tool and data hub system providing troubleshooting and condition monitoring solutions for electrical and mechanical systems from this supplier.

The ABB Ability™ Marine Remote Diagnostic System (RDS) comprises diagnostic elements installed onboard, remote connection hardware, and software including diagnostic solutions. Diagnostic data are sent to experts ashore for analysis and advice. Additionally, the remote diagnostic system helps engineers onboard to troubleshoot technical incidents for if the owner has signed the relevant contracts with ABB Marine Service Global Technical Support Center for 24/7 service. All diagnostic data are sent onshore by a secure network using secure ABB Remote Access Platform (RAP). An ABB service engineer may carry out remote diagnostic, maintenance or troubleshooting tasks only with prior crew permission.

The RDS concept comprises three main functionalities: Remote connection, Diagnostic solutions and Service agreements. Remote connection functionalities encompass Remote connection between ABB experts and the customer system, Secure SSL-encrypted communications (ABB RAP) through a satellite link and Automatic transfer of files to/from vessel and the ABB Ability™ Collaborative Operations Center having Service Level 2 for RDS Prevention and Service level 3 for RDS Prediction. Diagnostic solutions cover dedicated monitoring and diagnostics platform for collection, storage and analysis of data using modular platform with D4-diagnostic solutions. ABB Ability™ Marine Remote Diagnostic System (RDS) are achieved by co-operation with ABB Marine Service Global Technical Support Center after signing Service agreement for RDS Troubleshoot - Level 1, RDS Prevention - Level 2 or RDS Prediction - Level 3.

The Remote Diagnostic System simplifies the fault finding process and reduces the time required to identify and solve problems. Additionally, RDS enables condition monitoring and preventive maintenance of electrical and automation systems equipment, which helps to detect potential issues before they escalate, degrade performance or cause system failure. The concept of RDS is presented in Fig. 14.3.1.

The ABB Remote Diagnostic System can be used for diagnostics and monitoring of the following types of major equipment:

– Thruster—for Thrusters Motors, ABB VFD
– Anchor Winch—for Winch Motors, ABB VFD
– Medium Voltage Switchboard—ABB protection Relay, ABB Circuit Breaker, MSB Bus-bars
– Generators.

The RDS diagnostic system consists of the following subsystems:
D4Machines including AC500
Remote diagnostic services including condition monitoring of critical machines i.e. thruster motors, propulsion motors with AC500 hardware and dedicated software. AC500 is a Data Acquisition Units (DAU) that is suitable for any size of motor and generator. Each DAU has sixteen (16) channels for simultaneous high frequency vibration, current and voltage data, thirty two (32)

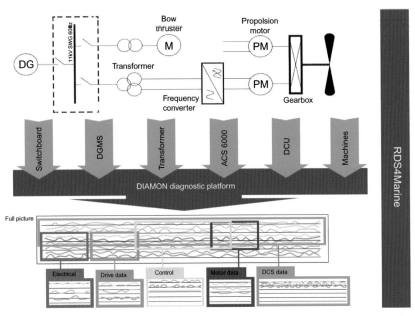

FIG. 14.3.1 RDS concept. *(Based on ABB webpage https://library.e.abb.com/public/97e02350b7 e6330bc1257c47004b1622/RDS%20Marine_Brochure%202014.pdf.)*

channels for temperature data. Rotation Speed and Voltage RMS derived from the Frequency Converter drive.

D4Switchboards including REM/REF 500-series, Relion Protection Relays
Remote diagnostic services for switchboards include diagnostic of:

– Medium Voltage Protection Relays i.e. alarm and events log history, circuit breaker opening/closing count, power generation and distribution monitoring, Transient recorders upload, Power Quality Analysis (e.g. Total Harmonic Distortion, Voltage Asymmetry, distortion etc);
– Measurements available for generators, transformers, thruster motors, bus transfer, measurement panels and any other feeders part of the MV switchboard.

D4Propulsion including ACS6000, ACS800, STADT, DCU drives
Remote diagnostic services for Propulsion include:

– Medium Voltage and Low Voltage Frequency Converters fault logging and handling, dataloggers upload, signals and parameters long term monitoring, Service Message for time based maintenance;
– Distributed Control Unit (DCU) application (i.e. alarms and events upload, continuous monitoring of critical signals, alarm and signals statistics).

D4Drilling including ACS800, ACS600, DDCA
D4Azipod including Azipod XO, Azipod V, Azipod C, for propulsion devices

Working principle is presented in Fig. 14.3.2.

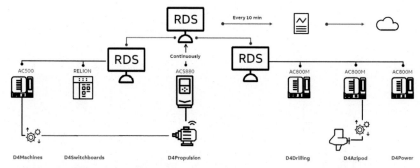

FIG. 14.3.2 RDS working principle.

Collected data from D4Machines AC500 controller, Relion relays, D4Propulsion AC5880 controller, D4Drilling AC800M controller, D4Azipod AC800M controller and D4Power AC800M controller are sent to ABB Marine Service Global Technical Support Center cloud every 10 minutes.

Example of RDS4Marine application is presented in Fig. 14.3.3.

An example ABB's Marine Remote Diagnostic System installed on DP3 Accommodation Unit consists of an RDS Centralized Cabinet containing ABB PPC, APC910 – D4Switchboard Module, APC910 – D4Machines Module and three (3) diagnostic network Switches: D4Winch system, D4Switchboard system and D4Thrusters system. The switches installed in the RDS4Marine Centralized Cabinet are connected to switches on the equipment as follows;

RDS D4Switchboard System—The central D4Switchboard Switch is connected by diagnostic network CAT6 to a switch installed in the No. 2 11kV Main Switchboard. Switches from Main Switchboard No. 1 and 3 are connected by diagnostic network CAT6 to Main Switchboard No 2. The switch in each Main Switchboard is connected to two (2) MOXA Nport 5400 Serial to Ethernet communication devices collecting diagnostic data from four (4) REF Protection Relays.

RDS D4Thrusters System—The central D4Thrusters Switch is connected by diagnostic network GOV type to Switches installed in Thrusters Drives No. 1, No. 2 and No. 6. A switch installed in Thruster Drive No 2 is connected via AC800M and PP8xx control network to a switch in Thruster Drive No 4, and a switch installed in Thruster Drive No 3 is connected via AC800M and PP8xx control network to a switch in Thruster Drive No 5.

The switch installed in each Thruster Drive is connected via a AC800M and PP8xx control network to its ACC800M Programmable Automation Controller and PP835 Operator Panel. A NDBU-95C Optical Driver installed in each Thruster Drive is connected to its Inverter Unit, Liquid Cooling Unit and Incomer section, and is connected additionally to a DM4000 Drive Monitor in the case of Thruster Drives No 1, 2, 5, 6.

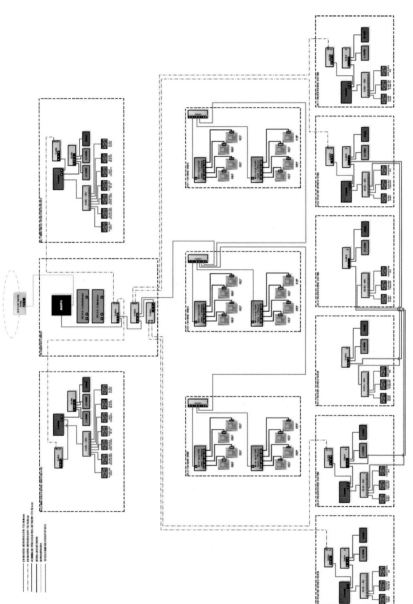

FIG. 14.3.3 Example of RDS application. (*Based on ABB documentation.*)

RDS D4Winch System—The central D4Winch Switch is connected via Diagnostic network GOV type to switches installed in the two (2) Winch Drive Switchboards. A separate switch installed in each Winch Drive Switchboard is connected via AC800M and PP8xx control network to two (2) AC800M Programmable Automation Controllers and one PP845 Operator Panel. Diagnostic Data from Inverter Units, Liquid Cooling Unit, Dynamic Brake Unit and Incomer Sections are sent via NDBU-95C Optical Driver to DM4000 Drive Monitor.

ABB developed dedicated software for RDS system consisting of RDS Software Platform, D4Switchboard Software, D4Winch Software, D4Machines Software, D4Propulsion Software and RAP/VSE Remote Access Platform Software for secure access from ABB Ability™ Collaborative Operations Centre to Vessel.

14.4   Wartsila remote diagnostic system

The Wartsila Integrated Automation System (WIAS) can be extended to provide remote diagnostic functions by adding a Virtual Private Network (VPN) connection to Wartsila specialists ashore. The Wartsila Remote Diagnostic System is presented in Fig. 14.4.1.

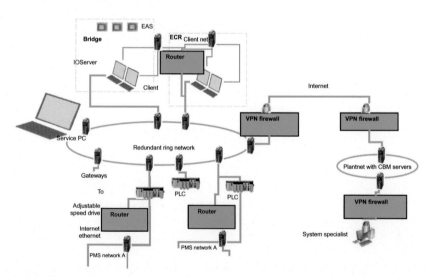

FIG. 14.4.1   Wartsila remote diagnostic system. *(Based on Wartsila WIAS General Functional Description.)*

Data collected onboard in I/O Server Operator Station, PMS 'A' and PMS 'B' Programmable Logic Controllers (PLCs) installed in Field Termination Cabinets (FTCs) are transferred via Routers, VPN network, VPN firewalls and internet to Wartsila servers ashore and are thus available to Wartsila specialists. For safety reasons each connection session between vessel and onshore Wartsila specialists must be approved by onboard vessel management.

Chapter 15

Control and monitoring systems certification

Chapter outline

Abstract

All equipment having an influence on safety of a vessel or offshore unit requires to be certified by the Classification society or an organisation recognized by the Maritime Administration. Exact certification requirements are included in Classification Societies rules for equipment Type Approval or equipment certification based on technical documentation approval and equipment tests supervised by classification society surveyor.

15.1 Classification society certification

Control and monitoring system installation and equipment certification requirements are determined by Classification Society rules. The certification process starts with approval of the system or equipment technical documentation, and a test programme after assessment to comply with the provisions of applicable rules and regulations. The classification Society issues the equipment certificate based on system or equipment positive test results.

A class society may issue a Type Approval Certificate valid for five years for the manufacturer of a system or equipment basing on approved technical documentation, positive tests results and satisfactory results of manufacturer's Quality Control system. Type Approval Certificates are issued for installations and equipment that have passed environmental tests described in Classification Society rules and International Association of Classification Societies (IACS)[1] Unified Requirements (UR) E10 Test Specification for Type Approval. This kind of supervision of standardised maritime equipment is called 'Indirect survey'. Based on a Type Approved certificate a manufacturer's technical department is authorised to issue a system or equipment certificate that is later verified

---

1. Based on webpage http://www.iacs.org.uk/

Ship and Mobile Offshore Unit Automation. https://doi.org/10.1016/B978-0-12-818723-4.00015-6

during classification survey of the vessel or offshore unit. No specific tests of the equipment are required at the vendor premises with the Type Approval method.

The Classification Society may supervise the installation of control and monitoring systems and equipment production during 'Direct survey'. In this case, the Classification Society surveyor will issue a certificate for equipment that is produced according to approved technical documentation and has passed tests in the presence of a class surveyor. The scope of such tests is smaller than type approval tests, but all features influencing vessel/equipment safety are verified and confirmed.

Certification requirements are determined by Classification Society rules.

15.2   Statutory certification

Control and monitoring systems equipment certification requirements are determined in statutory documents, i.e. SOLAS Convention, MODU Code, FSS Code[2]. SOLAS Convention and MODU Code determine regulations for Radio-communication equipment and FSS Code determines regulations for fixed fire detection and fire alarm systems. Equipment certificates are issued by the Administration or a Recognized Organization (RO) such as Underwriters Laboratory (UL) which is recognized by the flag administration, based on positive equipment test results.

---

2. FSS Code—International Code for Fire Safety Systems

Chapter 16

Control and monitoring systems installation

Chapter outline

Abstract

This chapter outlines requirements and guidance for installation of control and monitoring system. In this chapter are described typical documents used to perform installation of control and monitoring systems in controlled conditions. Such documents include shipyard standard practices, suppliers' Installation and Commissioning instructions. In addition installation guidelines for automation systems elements and Electromagnetic Compatibility considerations of equipment installed onboard are provided.

16.1 General

Control and monitoring systems are installed by shipyards, shipyard subcontractors or control and monitoring systems suppliers. Shipyard technology departments have developed Standard Practice documents for Electrical Installations that include control and monitoring systems. Such Standard Practice documents are developed according to Classification Societies rules, control and monitoring equipment vendor's installation instructions and the recommendations of electrical cable producers, as well as the shipyard's preferred way of working.

Standard Practices are agreed with the Owner of the vessel or offshore unit as part of the contract, and the shipyard and shipyard subcontractor's workers must then follow them when installing control and monitoring equipment and cabling. Control and monitoring systems suppliers prepare separate instructions for their equipment installation and commissioning. Standard practices are available for use by the shipyard workers and the quality control department.

16.2 Standard practices

Standard practices usually cover the subjects listed below, for which typical requirements are given:

– Cable installations—Cables to be routed to have adequate ventilation. Minimum distance between cables and heat radiating equipment shall be 500 mm. Cable ways shall not be installed below pipes connections. Cables shall be protected from mechanical damage by metal guards. Cables shall enter the equipment and junction boxes from the bottom or side. Instrumentation cables shall be routed separately from power cables to eliminate interference and line noise. Intrinsically Safe (IS) cables can run on the same instrument cable tray but shall be segregated by metal separator as per Classification Rules requirements. Multi Cable Transits (MCT) shall be installed when several cables are passing through watertight, gastight decks or bulkheads. All cables shall be marked at both ends with the cable reference numbers using flame retardant marks. Cables shall be fixed with clips maximum 300 mm apart. Cables shall run in continuous lengths between equipment and place where they are connected. Medium voltage cables, low voltage cables, instrument cables, Intrinsically Safe cables and VFD cables shall be segregated. Cable bending radius shall not be smaller than required by Classification Society rules and IEC standards—unarmoured/unbraided cable diameter up to 25 mm, minimum internal radius of bend 4D, over 25 mm radius 6D, composite polyester or metal laminate type screened cables radius 8D, for medium voltage single core cables radius 12D, and for 3-core cables radius 9D.
– Sketches of horizontal/vertical cable ladders/trays, segregated medium voltage cables, low voltage cables, instrument cables, Intrinsically Safe cables and VFD cable arrangements and horizontal/vertical spaces between them. Sketches of cable installations in machinery spaces, accommodation space corridors, cabins, cable installations in thermal insulation etc.,
– Sketches of cables typical tagging,
– Sketches of bulkhead and decks MCT cable penetrations montage and welding,
– Sketches and installation principles of equipment mounting in Engine Room and accommodation spaces for electrical, control and monitoring items such as switchboards and distribution boards, transformers, control panels, lighting fixtures, speakers, pushbuttons, smoke and heat detectors, escape indicators,
– Sketches and cables/wires termination principles using termination lugs,

– Instrumentation and communication cables shielding and termination principles,
– Cables and wires identification principles,
– Sketches of electrical, control and monitoring equipment connection and earthing methods including earth cross section size.

16.3   Installation and commissioning instructions

Control and Monitoring Systems producers/suppliers often develop specific instructions/specifications for their equipment installation and commissioning. Such a specification often includes methods of equipment earthing and connecting cables to control and monitoring equipment. In Figs. 16.4.1–16.4.4 are some example drawings included in installation instructions delivered to shipyard and Owner together with control and monitoring equipment.[1]

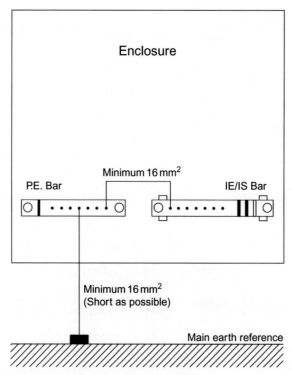

FIG. 16.3.1   Basic Kongsberg Maritime earth bar principle.

1. Extract from Kongsberg Specification of Earthing, Shielding and Power distribution principles.

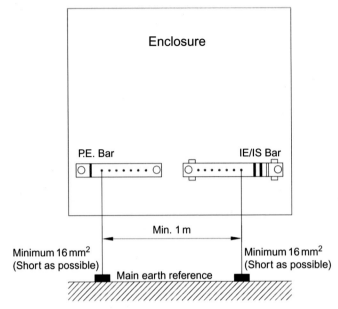

FIG. 16.3.2 NORSOK Earthing requirement.

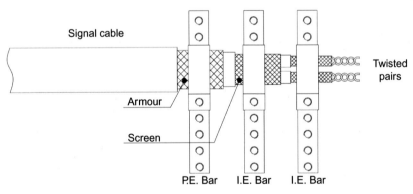

FIG. 16.3.3 Signal cable with armour termination.

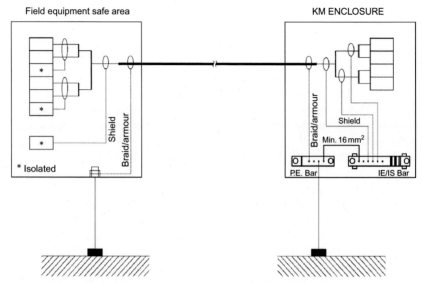

Field equipment safe area

KM ENCLOSURE

* Isolated

FIG. 16.3.4   Instrument cable with overall shields and inner (pair) shield

16.4   Control and monitoring systems in hazardous areas

Hazardous areas are all those areas where, due to the possible presence of a flammable atmosphere arising from the battery/paint fumes or drilling or hydrocarbon production operations, the use without proper consideration of machinery or electrical equipment may lead to fire hazard or explosion.[2]

For each vessel/offshore unit a Hazardous Area plan is developed, showing the potentially hazardous area layout, and also including a list of the electrical equipment installed in hazardous areas, and their locations. Such plan must be approved by the vessel/offshore unit flag administration and/or the Classification Society representing the administration.

Hazardous areas are classified in the International Electro-technical Commission (IEC) standard series IEC 60079[3] as below:

- **Zone 0**—place in which an explosive atmosphere consisting of a mixture with air of flammable substances in the form of gas or vapour is present continuously or for long periods or frequently;
- **Zone 1**—place in which an explosive atmosphere consisting of a mixture with air of flammable substances in the form of gas or vapour is likely to occur in normal operation occasionally;
- **Zone 2**—place in which an explosive atmosphere consisting of a mixture with air of flammable substances in the form of gas or vapour is not likely to occur in normal operation but, if it does occur, will persist for a short period only.

2. Code for the Construction and Equipment of Mobile Offshore Drilling Units (MODU Code).
3. IEC web page http://www.iec.ch/standardsdev/publications/is.htm.

Similar definitions of Hazardous areas are provided in classification societies rules, IMO MODU Code, and IMO SOLAS Convention. Care should be taken in mixing different definitions within one project, as different definitions exist.

For ships, the SOLAS definition provides the following examples;

Zone 0 examples are the interior of a cargo tank on a crude oil, oil products, or chemical products tanker carrying flammable liquids (other than liquefied gases) having a flash point not exceeding 60 °C. In case of liquefied gases, the cargo tank itself and the surrounding secondary barrier spaces, the pump room are classified as zone 0.

Zone 1 examples include spaces adjacent to and below the tops of cargo tanks carrying crude oil, oil products or chemicals etc. with flash point up to 60° C, separated by a single deck or bulkhead from Zone 0, cargo pump rooms, and spaces where pipes for the above cargoes are passing through. This definition covers ballast tanks adjacent to cargo tanks.

Equipment to be installed in a hazardous area needs to be certified for its particular installation location.

In Zone 0, electrical equipment intended to be installed must be Group II, Category 1G with protective concept Ex ia—Intrinsic safety with limitations on energy as well as arcs and temperature.

Intrinsically safe protection relies on limiting the amount of electrical energy available to a circuit under normal or abnormal conditions to a level too low to ignite a flammable atmosphere. The parameters to be considered include thermal energy (hot surfaces) and sparking.

Intrinsic safety equipment is equipment isolated by a barrier unit; the barrier limits the energy in the hazardous area to the extent that it cannot cause a spark which could start an explosion. The barrier is normally located in a safe area. The cable from the barrier unit to the intrinsically safe unit in the hazardous area must be routed separately from other, not intrinsically safe cables to prevent the cable picking up additional power through induction that would exceed the IS safe limit.

In Zone 1, electrical equipment may be installed which is classified for Zone 0 (as above), or equipment Group II, Category 2G with following protective concepts:

- Ex d—Flame proof, where propagation of an explosion from inside the equipment to the outside is excluded or
- Ex e—Increased safety, to avoid of arcs, sparks and excessive temperature or
- Ex ib—Intrinsic safety, with imitation of energy as well as arcs and temperature or
- Ex p—Pressurization, where external explosive atmosphere is kept at a distance from the ignition source or
- Ex o—Oil immersion, where external explosive atmosphere is kept at a distance from the ignition source or
- Ex q—Powder filling, where propagation of an explosion from inside the equipment to the outside is excluded or

– Ex m—Encapsulation, where explosive atmosphere kept at a distance from the ignition source.

In Zone 2, electrical equipment may be installed which is classified for Zone 0 and 1 (as above), or equipment Group II, Category 3G with following protective concepts:

– Ex ic—Intrinsic safety, with limitation of energy as well as arcs and temperature;
– Ex n—'n' type of protection allowable for Zone 2.

To harmonise the certification system for Ex equipment and have one single certificate recognised and accepted worldwide, the IEC established the IECEx System that includes requirements and procedures for IECEx certification.

The European Union has unified requirements for ships and onshore electrical equipment installed in hazardous areas by publishing the ATEX Directive. This is also applied on UK offshore installations.

It should be noted that in North America there are different standards for division and zone classification, for example National Electrical Code (NEC) in USA, Canadian Electrical Code (CEC) and American Petroleum Institute (API).

In practice on ships, hazardous areas include:

– hydrocarbon cargo tanks (for liquids with flash point below $60^{\circ}C$);
– hydrocarbon cargo pump rooms;
– ballast tanks adjacent to hydrocarbon cargo tanks;
– 3 m high area above deck above such cargo tanks;
– battery rooms, for which installed electrical equipment shall be minimum group II C and temperature class T1;
– paint stores, for which installed electrical equipment shall be minimum gas group II B and temperature class T3;
– welding gas bottle stores, for which installed electrical equipment shall be minimum gas group II C and temperature class T2;
– refrigerating systems containing Ammonia ($NH3$), for which installed electrical equipment shall be minimum gas group II A and temperature class T1.

16.5   Electromagnetic Compatibility (EMC) considerations

Control and Monitoring equipment and systems installed on ships and offshore units are vulnerable to electromagnetic interferences. Potential interference sources are:

– Natural—atmospheric, electrostatic discharge, or
– Technical—Intentional as Radio Frequency generation i.e. Radiocommunication, telephones or

- Intentional—generated by power electronics, energy distribution systems, ignition systems, commutating electric motors etc.

Vessel owners/operators request shipyards or their subcontractors to fulfil the requirements of the IEC standard (IEC 60533) related to electrical and electronic installations in ships—Electromagnetic Compatibility (EMC)—Ships with a Metallic Hull, the European Union Directive of the European Parliament and of the Council on the harmonisation of the laws of the Member states relating to the making available on the market of electrical equipment designed for use within certain voltage limits, and EU Directive on the harmonisation of the laws of the Member States relating to electromagnetic compatibility.

IEC Standards specify limit values for control and monitoring systems equipment immunity and electrical equipment emission limits, i.e. electrostatic discharge, electromagnetic field, conducted radio frequency, Burst/Fast transients, surge voltage, radiated emission, conducted emission etc.

To reduce EMC influence on control and monitoring systems the following counter measures are implemented:

- Grounding or Earthing audio equipment,
- Using shielded cables where the signal wires are surrounded by an outer shield that is grounded at one or both ends,
- Using shielded housings—a conductive metal housing with an interference shield and cables are connected through metal gaskets,
- Using harmonic wave filters or filtering systems with RF chokes or RC elements,
- Using transient absorbers,
- Installing sensitive control and monitoring equipment at a safe distance from frequency converters and their cables,
- Installing frequency converters as close as possible to the motor they serve.

Control and monitoring equipment is tested and certified according to classification societies requirements and IEC standards requirements. Where required, equipment is marked 'CE' according to European Union Directives. Equipment fulfilling all EMC requirements is able to operate in their electromagnetic environment without any fault, and the same equipment does not influence the electromagnetic environment to the extent that functions of other equipment installed on vessel or offshore unit are affected.

Chapter 17

Mechanical completion and commissioning of control and monitoring systems

Chapter outline

Abstract

This chapter outlines procedural requirements for commissioning control and monitoring systems including mechanical completion phases e.g. documentation, certificates, check records, punch lists and handover.

17.1 Mechanical completion and commissioning of control and monitoring systems

For offshore projects, especially in Norway, Mechanical Completion and commissioning of control and monitoring systems is often carried out according to NORSOK Standard (Z-007). Common Requirements, Mechanical Completion and Commissioning.[1] This standard includes:

- Completion principles—Overall goal to verify that system is designed and built to fulfil its purpose and specified requirements;
- Mechanical Completion (MC)—Mechanical Completion phases, documentation, certificates, check records, punch list register, handover, typical: mechanical, electrical and instrument/telecommunication completion activities. Sometimes the Mechanical Completion process is called Pre-commissioning;
- Commissioning—Commissioning phases, Commissioning procedure content: objective, description, list of temporary equipment and consumables, Health/Environment/Safety issues, preservation, commissioning scope, planning and handover;

---

1. Based on webpage https://www.standard.no/en/sectors/energi-og-klima/petroleum/norsok-standards/#.WpgHzmepXYU.

Ship and Mobile Offshore Unit Automation. https://doi.org/10.1016/B978-0-12-818723-4.00017-X

The shipyard usually prepares commissioning procedures for the control and monitoring systems. Typically, such procedure includes the following chapters:

- General—describing needs for management of work by Permit to Work system, commissioning personnel knowledge and training, listing safety procedures and shipyard Standard Practices,
- Purpose—to ensure proper system installation and to demonstrate the operational capabilities by function test,
- References—containing system operation philosophy, functional description, installation drawings,
- Pre-commissioning checks—to be verified and status noted: equipment arrangement according to relevant drawings, cables connections, name plates, loop tests results (if applicable), software pre-commissioning results, Factory Acceptance Test (FAT) results,
- Test procedures—to be tested and status noted: commissioned equipment hardware, software, power supply test, controlled and monitored equipment tests, power distribution to controlled and monitored equipment
- Punch list—describing control and monitoring system test non-conformances.

Control and monitoring systems Mechanical Completion (MC) is a final stage of construction activities to verify systems completeness (e.g. proper cable connection and termination) and demonstrate positive test results according to the project specification.

One of Engine Room (ER) on Save Boreas Accommodation Rig after mechanical completion is in Fig. 17.1.1.

© Prosafe

FIG. 17.1.1 *Safe Boreas* accommodation rig engine room. *(Picture received from Prosafe AS, Norway.)*

Control and monitoring systems commissioning follows the construction tests (mechanical completion) to test the functioning of the subsystems and the complete system in conditions as close as possible to normal intended operational conditions. In the case of power management systems tests will be carried out to prove generator load reaction and start/stop functions and blackout recovery. In the case of offshore units with ESD systems, various shutdowns shall be simulated and the recovery of the system from these events examined.

Chapter 18

Control and monitoring systems maintenance

Chapter outline

Abstract

This chapter outlines different systems to supervise maintenance of control and monitoring system elements used by industry, for example Planned Maintenance Systems (PMS), and Computerized Maintenance Management System (CMMS).

18.1 Control and monitoring systems maintenance

Sophisticated vessel and offshore units control and monitoring systems are equipped with modern sensors i.e. pressure transmitters, temperature transmitters, pressure switches, thermostats and gas, smoke, temperature detectors. All those systems and equipment need maintenance. Usually owners and operators develop their maintenance strategy outlining the duties and responsibilities for control and monitoring equipment maintenance and defining the applicable rules, regulations and maintenance procedures.

Due to the very large quantity of control and monitoring devices on ships and offshore units the owners, vessels managers and their technical departments are often using computerised Information and Planning Systems (IPS) and Planned Maintenance System (PMS), sometimes described as Computerised Maintenance Management Systems (CMMS). Such systems help them to operate the overall fleet or the individual vessel in a safe and efficient manner and to comply with industry regulations, statutory requirements and classification societies rules. A Planned Maintenance System helps the personnel onboard in planning ship maintenance, reporting work done, keeping track of stock and handling of spare parts and supplies.

The Planned Maintenance System is a licensed software used on board the vessel and in the owners/operator's onshore technical department which can audit reported maintenance activities remotely and order preventive actions if necessary. Such a system is designed to provide ready access to all necessary maintenance information i.e. ship/offshore unit name, system name, equipment

Ship and Mobile Offshore Unit Automation. https://doi.org/10.1016/B978-0-12-818723-4.00018-1
163

name, maintenance/work instruction or work order (WO), maintenance planning-schedules, resources, work report (WR), work done, maintenance history, guarantee claims, accident/near miss reporting, failure reporting, budgets, spares inventory and technical department maintenance engineer's recommendations. It is possible to print Work Orders (WO) and filtered maintenance schedules and Work Instructions (WI) from the system. All records in the PMS are identifiable.

To reduce off-hire costs and downtime, sometimes the vessel or offshore unit managers hire specialised companies which are certified by class to provide the necessary maintenance services.

Maintenance activities are a necessary condition of maintaining vessel classification, and are part of the checks performed by the Classification Society at Annual Survey, Renewal Survey and can also be part of a Continuous Machinery Survey (CMS) or Planned Maintenance Scheme (PMS) based on actual working hours and equipment vendors' recommendation.

Chapter 19

Vessel/offshore unit certification

Chapter outline

Abstract

This chapter outlines requirements for certification of control and monitoring systems and equipment installed on vessel or offshore units. Additionally there are outlined Health and Safety Executive rules and guidance applicable on United Kingdom continental shelf (UKCS), Norwegian Maritime Authority (NMA) requirements guidance published by Petroleum Safety Authority (PSA) and International Marine Contractors Association (IMCA) guidelines.

19.1 Classification society certification

At the Owner's request, a Classification Society assigns a class to the vessel or offshore unit that fulfils classification rules requirements and regulations, including the necessity of having certified control and monitoring systems and installations. After a positive result of classification survey, the Classification Society issues the Class Certificate and Machinery Certificate.

19.2 Statutory certification

Statutory IMO certificates for ships include; Safety Certificate for Cargo Ships, Safety Construction Certificate for Cargo Ships, Safety Equipment Certificate for Cargo Ships, Safety Radio Certificate. For Offshore units, the equivalent of these is the IMO Mobile Offshore Drilling Unit Safety Certificate which is issued after positive statutory survey, including requirements of equipment statutory certification.

Ship and Mobile Offshore Unit Automation. https://doi.org/10.1016/B978-0-12-818723-4.00019-3

19.3 Offshore unit coastal state approval

19.3.1 Health and safety executive (United Kingdom)

The Health and Safety Executive (HSE) develops and enforces regulations relating to health and safety on offshore installations and offshore units working within the United Kingdom continental shelf (UKCS) and to oil and gas operations in the territorial waters adjacent to Great Britain. The HSE publishes Health and Safety Guidance (HSG) documents, and Health and Safety Regulations.

The governing regulations requiring management of Major Accident Hazards (MAHs) are in The Offshore Installations (Offshore Safety Directive) (Safety Case etc.) Regulations 2015. HSE has published the Offshore Installations (Offshore Safety Directive) (Safety Case etc.) Regulations 2015 Guidance on Regulations which is a very useful document for Mobile Offshore Unit owners and operators.

Also relevant for control and monitoring systems are 'The Electricity at Work Regulations' and 'Prevention of Fire and Explosion, and Emergency Response on Offshore Installations (PFEER) and the related PFEER Approved Code of Practice and Guidance'.

HSE staff carry out offshore surveys to assess and reduce risks for health and safety. When assessment results are unsatisfactory the HSE Officer prepares Notifications or Non-Acceptance Issue Note(s) listing all deficiencies to be dealt with.

19.3.2 Petroleum safety authority and the Norwegian maritime authority

Mobile Offshore Units intended for use in petroleum activities offshore Norway are required to fulfil Norwegian statutory requirements and guidelines published by the Petroleum Safety Authority (PSA) and the Norwegian Maritime Authority (NMA). These requirements are additional to international statutory instruments such as IMO MODU Code, and the rules of Classification Societies such as American Bureau of Shipping (ABS), Det Norske Veritas Germanischer Lloyds (DnVGL) and Lloyds Register (LR).

Mobile Offshore Units intended to work on the Norwegian Continental Shelf are required to obtain an Acknowledge of Compliance (AoC) document. The Norwegian Shipowners Association has published a 'Handbook for Application for Acknowledgement of Compliance (AoC) describing Mobile Offshore Units verification methods: inspections and surveys, use of class and maritime certificates, verification during performance of maintenance analysis and evaluations.

19.4   Client and IMCA audits of control and monitoring systems

An offshore unit may be audited by the end user client or consultants appointed on his behalf. For control and monitoring systems the scope of such audits usually covers DP systems and thruster control systems. One of the widely recognised standards for auditing and surveying offshore units is provided by the International Marine Contractors Association (IMCA). IMCA publishes the following publications[1]:

– Guidelines for The Design and Operation of Dynamically Positioned Vessels (M 103);
– Common Marine Inspection Document (M149);
– Guidance on Failure Modes and Effect Analysis (FMEA) (M166);
– A guide to DP electrical power and control systems (M206);
– Guidance on satellite-based positioning systems for offshore applications (M242);
– Guidance on vessel USBL systems for use in offshore survey, positioning and DP operations (M244);
– Guidance on mobilisation requirements for offshore survey operations (S016);
– Guidance on satellite-based positioning system for offshore applications (S024) etc.

Client audit and survey findings are included in the Report that is delivered to Offshore Unit Owner or Operator to carry out corrective or preventive actions before starting new contract.

---

1. https://www.imca-int.com/publications/.

Chapter 20

Control and monitoring systems evolution and future

Chapter outline

Abstract

This chapter describes the evolution of control and monitoring systems from the 20th century to the present date. Also described are future trends automation systems including trends to design and build of autonomous ships. Also described are trends in development of Artificial Intelligence techniques in Integrated Automation Systems (IAS), Dynamic Positioning (DP) Systems, Position Mooring (PM) Systems and autonomous ships development.

20.1 Control and monitoring systems evolution and future

Development of control and monitoring systems on board vessels began in the early 1950s. At that time an Engine Room Logbook was simply kept on a desk in the Engine Room near the main switchboard and was regularly updated by the Marine Engineer Watchkeeper and countersigned by the Chief Engineer. Control systems were mainly limited to remote control of main propulsion, automatic control of hydrophore pumps, air compressors etc. Later, the Chief Engineer desk was located in a small room that was separated from the Engine Room. Near the desk were installed some monitoring instruments.

By the beginning of the 1960s most new ships had an Engine Control Room (ECR) containing the Engine Control Room Console (ECRC) as well as the engineer's desk. Development of electric, electronic, pneumatic and hydraulic control systems allowed the control and monitoring of most Engine Room equipment from the ECRC. In the 1970s, development of programmable digital computers allowed the control and monitoring of machinery systems with Supervisory Control and Data Acquisition (SCADA) systems.

In the 21st Century, ships and offshore units are fitted with Integrated Control Systems (ICS) including Integrated Automation Systems (IAS), Dynamic Positioning (DP) systems, Thruster Control Systems, and Safety Systems with

Ship and Mobile Offshore Unit Automation. https://doi.org/10.1016/B978-0-12-818723-4.00020-X

Emergency Shutdown (ESD) and Fire & Gas functionality. Modern Integrated Control Systems are based on distributed processing and data acquisition units connected by redundant networks - either fibre optic cables or twisted pair cables. Main system elements are redundant, including Human Machine Interfaces (HMI) - operator stations, monitors, process controllers with dual computers, Input/Output (I/O) modules and Uninterruptable Power Suppliers (UPS).

The number of digital and analogue I/O for control and monitoring systems varies greatly depending on the vessel type and size, but is typically between 5000 to 9000. Laser printers are used for data logging. Monitors on Operator Stations provide system mimic diagrams illustrating process flows, existing alarms and previous alarms and events list. The number of mimic diagrams (a.k.a. process views) can be about 150. Built-in self-diagnostic facilities monitor the system components and communication links. Fail-safe mechanisms are automatically activated in case of failure.

Modern Integrated Control Systems software is built up from algorithms precisely defining different control and monitoring equipment operation sequences. Some Dynamic Positioning systems use a vessel mathematical model—Kalman Filter—for calculations of vessel positioning and heading response.

A potential next step in vessels and offshore unit control systems development is fully or partly autonomous ships. There are extremely high requirements for systems to be installed on such vessels.

Currently engineers and scientists are working on development of Maritime Autonomous Surface Ships (MASS) that can operate partially or fully independently of human interaction. Such systems require to accommodate new technologies including the highest level of digitalization and automation having in mind vessel safety. During design and development processes different degrees of autonomy will be taken into consideration, i.e.

- Ship with automated processes and decision support in which some crew is on board to operate and control shipboard systems, and some operations are automated
- hip is remotely controlled and operated from another location but limited number of crew members are onboard
- Ship is fully remotely controlled and there is no crew onboard
- Ship is fully autonomous and all operating systems make decisions and determine actions by themselves

Maritime Autonomous Surface Ships (MASS) will require installation of advanced sensors which take care of the lookout duties onboard the vessel by continuously fusing sensor data from navigational systems such as Radio Detecting And Ranging - RADARs, Automatic Identification System (AIS) combined with daylight and infrared camera images. Autonomous ships additionally require designated autonomous navigation systems that follows predefined voyage plans with certain degrees of freedom to adjust the route having in mind weather changes and navigational situations.

Control and monitoring systems on autonomous ships will require to be designed to fulfil ship's engine room and propulsion automation systems advanced failure pre-detection functionalities having in mind vessel optimal efficiency and propulsion redundancy.

There will be additional challenges for interfacing autonomous ships systems with shore based personnel including: Shore Control Centre Operator who monitors the safe operation of several ships simultaneously from shore station and controls the vessels by giving high level commands for example "change voyage plan", Shore Control Centre Engineer who assists the operator with any technical issues including maintenance plan based on condition based maintenance system program to ensure sufficient vessels technical systems reliability, and Shore Control Centre Situation Team that is able to take over direct remote of the vessel to ensure an appropriate situation awareness.

Maritime Autonomous Surface Ships (MASS) will require development of ship-shore and ship-ship communication systems to ensure security and reliability of integrated ship data networks.

Future autonomous ships will need special Advanced Sensor modules for objects detection and classification. Such systems will use input data from infrared and visual spectrum cameras, Radar, and Automatic Identification System (AIS) data to detect any surrounding objects and to determine if they may cause danger for the ship or need to be further investigated i.e. life rafts or other floating objects. This system shall collect all data from navigational, metrological and safety sensors to identify objects causing potential hazard for the ship.

Control and monitoring systems on autonomous ships need to be more advanced than on today's ships by adding sophisticated condition monitoring functions to prevent vessel critical systems malfunctions and breakdowns during the sea voyage. These systems need to be linked with maintenance planning systems. Control and monitoring systems shall deliver information necessary to monitor equipment diagnostics, for example equipment thermal overloads or vibration. This information will be used by decision support system in Shore Control Centres. Extended control and monitoring system data shall be aggregated to minimise satellite communication bandwidth.

Development of control and monitoring systems for Maritime Autonomous Surface Ships (MASS) will have influence on designing Integrated Automation Systems (IAS) for vessels with unmanned Engine Room and automation systems, dynamic positioning and position mooring.

All ship and offshore unit functions are based on condition detection, condition analysis, action planning and action control. To carry out a ship functions, new reliable vessel/offshore unit condition detection sensors must be employed, i.e. for example daylight cameras of different type as stereo or multispectral, infrared cameras and Light Detection and Ranging Cameras (LIDAR). These detectors shall be reliable in adverse conditions such as heavy seas, darkness, fog, heavy seas and snowfall. Autonomous ships need redundant Global Navigation Satellite System (GNSS) i.e. to existing USA Global Positioning

System (GPS) another systems need to be used, for example European Union Global Navigation Satellite System Galileo, Russian Global Navigation Satellite System (GLONAS) or China BeiDou Navigation Satellite System (BDS). Additionally Maritime Autonomous Surface Ships (MASS) need reliable nautical charts identifying Zones of Confidence (ZOC) and risks of collision and grounding. Supplementary sensors will be required to assess the capability of propulsion and steering at defined time as well as for predicting any possible changes in this capability.

Installed sensors will need to have homogenous redundancy by installing two or more sensors or heterogeneous redundancy by installing several sensors measuring different quantities and calculating the quantity without taking into account unexpected measurement. Condition analysis needs to be carried out by powerful computers when all relevant information is detected/measured. When processes conditions are determined then based on Artificial Intelligence computers develop an action plan and assign action control to assigned automated systems or onboard personnel.

In the future, the quantity of control and monitoring data to be handled is expected to continue to grow and new Decision Support Systems (DSS) based on Artificial Intelligence (AI) techniques may start to be used on board vessels and offshore units. To increase safety, many sophisticated computer based systems are already installed on board vessels and offshore units, but a large percentage of sea accidents and collisions are the result of human errors arising from difficulties to make proper decisions under stress and having to process large amounts of information and alarms. Such Decision Support Systems will be based on Artificial Intelligence using expert knowledge and operating on neutral network principles.

Artificial Intelligence technology requires development of AI-models that will help predict the most efficient way to operate a particular vessel or offshore unit automation system. One example of implementing Artificial Intelligence technology is 'Control system to reduce vessel fuel consumption' using AI simulations in different scenario before suggesting the most optimal route and performance setup taking into consideration a number of variables such as currents, weather conditions, shallow water and speed through water. Another example of using Artificial Intelligence technology to increase safe of navigation the vessel is development officer on watch 'adviser' that analyse data from shipboard sensors, cameras and microphones to create a picture of the hazards around the ship. Such picture may be presented on officer on watch portable tablet or Smartphone to help in safe navigation. In short time Artificial Intelligence will be used to reduce maintenance costs. Nowadays a Condition Based Monitoring (CBM) database is used to plan Condition Based Maintenance activities. Artificial Intelligence (AI) technology by deep learning capacity to analyse very large amount of high dimensional data will change Preventive Maintenance System (PMS) to higher level predictive maintenance system that is particularly valuable in the case of hard worked assets such as

Main Engines, by combining engine model data, maintenance history, data from sensors around engines such as sound, temperature etc., engine sensors data as lube oil temperature, cooling water temperature, vibration etc. and images from video cameras. Artificial Intelligence technology will reduce risks for workers in hydrocarbon exploration and production by replacing them with intelligent robots. Additionally, these robots will be used in oil fields exploration seep detection and so reduce sea pollution.

In the future, on board Artificial Intelligence systems will be operated with a common operating system to enable the connection of various types of equipment in the ship to one common operating control system. For example, using such a system crew may be able to read the data and control equipment using special applications in their Smartphone.

Before implementing above developments, all related cyber security threats, legal and liability matters must first be solved, and these are significant challenges in their own right. Cyber attacks may include remote or physical interaction with the vessel/offshore unit IAS or stealing/damaging control and monitoring sensors and actuators. Consequences of cyber attacks on fully autonomous ships could be extremely high because it may be not possible to correct the vessel situation by a crew.

Appendix 1A

List of contract regulations, requirements and standards usually included in ship Technical Specification[1]

Classification rules

- DNV-GL Rules for Classification, Ships—*applicable for automation and control monitoring systems: Part 4 Systems and components, Chapter 8 Electrical installations, Chapter 9 Control and monitoring systems and Chapter 11 Fire safety*

National rules & regulations

- Relevant Regulations and guidelines as set by the Maritime Administration.
- *Wheelmark—European Union's Council Directive 2014-90 EU (Marine Equipment Directive).*
- EU Directive 03/25/EC—Specific stability requirements for ro-ro passenger ships (Stockholm Agreement)
- Regulation (EU) 1257—2013 on Ship Recycling. Helsinki Convention—Convention on the Protection of the Marine Environment of the Baltic Sea Area
- EU Directive 1999/32/EC amended by 2005/33/EC with regard to operation within European territory
- EN 81-20 and EN 81-50 requirements for passenger and goods lifts
- Baltic Memorandum of Understanding for transport of dangerous goods in Ro-Ro ships

---

1. Contract regulations, rules and standards applicable to automation, control and monitoring are marked in italic text.

- EU Directive 2005/33/EC requirement regarding Use of Low Sulphur Fuel On-board Ships
- *CE marked as relevant for electrical equipment*

International rules & regulations

IMO conventions

- *SOLAS—International Convention for the Safety of Life at Sea—consolidated as amended*
- Load Line—International Convention on Load Lines—consolidated Edition 2005 as amended
- Tonnage—International Convention on Tonnage Measurements of Ships 1969—as amended
- International Regulations for Preventing Collisions at Sea, 1972, as amended
- *MARPOL—International Convention for the Prevention of Pollution from Ships—consolidated Edition 2011 as amended*
- COLREG—Convention to the International Regulations for preventing Collision at Sea 1972—consolidated Edition 2003 as amended
- *BWM 2004—International Convention for the Control and Management of Ships Ballast Water and Sediments, consolidated Edition 2004 as amended*
- GMDSS—Global Maritime Distress and Safety System 1992—as amended
- INMARSAT—International Mobile Satellite Organization (IMSO) 2008—as amended
- AFS—International Convention on the Control of Harmful Anti-Fouling Systems on Ships, 2001

IMO codes

- IS Code—International Code on Intact Stability—consolidated Edition 2008 as amended
- *FSS Code—Fire Safety System Code—consolidated Edition 2015 as amended*
- LSA Code—International Life-Saving Appliance Code—Resolution MSC. 48(66)
- ISPS Code—International Ship and Port Facility Security Code—consolidated Edition 2003 as amended

Other technical requirements

- PSPC Code—Performance Standards for Protective Coating Code—consolidated Edition 2006 as amended

- Code on Noise Levels on board Ships—consolidated Edition 2014 as amended
- *Code on Alerts and Indicators—consolidated Edition 2009 as amended*
- IMDG Code—International Maritime Dangerous Goods Code 2018
- Noise Levels—MSC.337(91) Code on Noise Levels on Board Ships
- Maritime Labour Convention (MLC), 2006 (Part A & B)

IMO resolutions

- Resolution A.601(15)—Provision and Display of Manoeuvring Information on board Ships
- Resolution A.581(14)—Guidelines for Securing Arrangement for the Transport of Road Vehicles on RoRo Ships
- *Resolution MSC.137(76)—Standards for Ship Manoeuvrability*
- *Resolution MEPC.107(49)—Revised Guidelines and Specifications for Pollution Prevention Equipment for Machinery Space Bilges of Ships*
- *Resolution MEPC.227(64)—2012 Guidelines on Implementation of Effluent Standards and Performance Tests for Sewage Treatment Plants*
- IMO MSC.404(96), Evacuation Analysis for Passenger Ship
- IMO MSC.409(97), Protection Against Noise
- IMO MSC.409(97), Foam-type Extinguisher for Boiler Space
- Resolution MSC.421(98), SOLAS 2020 Subdivision and Damage Stability
- Resolution MSC.421(98), Fire Integrity of Windows
- Resolution MSC.421(98), Requirements Applied to Vehicle Space
- MSC Circ.917, Guidelines on Fire Safety Construction in Accommodation Areas
- Resolution MSC.232(82)—Revised Performance Standards for Electronics Chart Display and Information Systems (ECDIS)
- *Resolution A.861(20)—Performance Standards for ship borne Voyage Data Recorders Amended by Resolution MSC.214(81)*
- *Resolution A.708(17) Navigation Bridge Visibility and Function, as may be superseded by navigational class notation*
- *MSC Circ.648 Guidelines for the operation, inspection and maintenance of ship sewage systems*
- *IMO Res.MEPC.269(68) Guidelines for Development of the Inventory of Hazardous Materials*
- *IMO MSC/Circ.1053 Explanatory Notes to the Standards for Ship Manoeuvrability*
- *IMO MSC/Circ.982 Guidelines on Ergonomic Criteria for Bridge Equipment and Layout*
- *IMO Res. A.1045(27) Referring to Pilot Boarding Requirements*
- *IMO Res. MSC.323(89) Adoption of Amendments to the revised Recommendation on Testing of Life-Saving Appliances*

- IMO MSC Res.192(79) *"Adoption of the Revised Performance Standards for Radar Equipment"*
- IMO MSC Res.334(90) *"Adoption of Amendments to the Performance Standards for Devices to Measure and Indicate Speed and Distance"*
- IMO Res. MSC.128 (75) *"Performance Standards for a Bridge Navigational Watch Alarm System (BNWAS)"* and MSC 282(86)
- IMO Publication No.978E *Performance Standards for Navigational Equipment (2011 Edition)*
- IMO MEPC 1 Circ. 642—*revised guidelines for handling oily wastes in machinery spaces of ships (integrated bilge treatment systems)*
- International Code for Application of the Fire Test Procedures, 2010, Res. MSC.307(88)
- International code for ships Operating in polar waters MSC.385(94) 10/201, as amended

Special rules & regulations

- ISO 3046-2002-1—Reciprocating internal Combustion Engines-Performance—Part1: Declarations of Power, Fuel and Lubricating Oil Consumptions and Test Methods
- ISO 8217-2012 Petroleum Products—Fuels (class F)—Specification of Marine Fuels
- ISO 8861-1998—Engine Room Ventilation in Diesel-Engine Ships—Design Requirements and Basis of Calculation
- International Telecommunication and Radio Regulation, 1973/1974/1976
- *IEC 60529—Degrees of protection provided by enclosures (IP Code)*
- *IEC 60092—Electrical installations in ships*
- *IEC 61000—EMC standards*
- ISO 9001—Quality Management System
- ISO 14001—Environmental Management System
- ISO 6954: 2000(E) Mechanical Vibration-Guidelines for Measurement, Reporting and Evaluation of Vibration with regard to Habitability on Passenger and Merchant Ships
- ISO 15016: 2015(E) Ships and Marine Technology-Guidelines for the Assessment of Speed and Power Performance by Analysis of Speed Trial Data
- *International Electro-technical Commission (IEC) publication 60533 Electrical and Electronic Installation on Ships-Electromagnetic Compatibility (1999)*
- ISO 14276-1 Ships and Marine Technology—Identification colors for the content of piping systems—Part 1: Main colors and media & Part2: Additional colors for different media and/or functions

- ISO 7547:2002 Ship and Marine Technology-Air conditioning and ventilation of accommodation spaces-design conditions and basis of calculation
- *IEC 60945 (2002/2008)—Maritime navigation and radio communication. Equipment and systems. General requirements*
- *IEC 62040 Series. (2008)—General and safety requirements for UPS. Uninterruptible power systems (UPS)*
- *IEC 60947 Series—Low voltage*
- *IEC 60470 and IEC 60056 for high voltage equipment*
- *IEC 60331/60332 Series—Tests for electric cables under fire conditions. Flameretardation of individual cables. Flame-retardation of cable bunches. Fire-resistant cables.*
- *IEC 60529—Degrees of protection provided by enclosures (IP Code). of protection provided by enclosures (IP Code)*
- *IEC 61000—EMC standards*
- IEC 60309 Series—Plugs, socket-outlets and couplers for industrial purposes
- IEC 62676-1-1 Series (2013)—Video surveillance systems for use in security applications. System requirements—General

Appendix 1B

List of contract regulations, requirements and standards usually included in offshore Unit Technical Specification[1]

Classification rules

1. *DNVGL-RU-OU-101 Rules for Classification Offshore drilling and support units*
2. *DNVGL-SI-0167 Verification for compliance with United Kingdom shelf regulations*

Regulatory requirements

1. IMO Code for the Construction and Equipment of Mobile Offshore Drilling Units (IMO MODU Code 2009) with latest amendments
2. IMO Resolution A.1005 (25) and MEPC 59—International Convention for Control and Management of Ship Ballast Water and Sediments
3. Application of the BWM Convention to Mobile Offshore UNITs—BWM 2/Circ. 46, 31st May 2013
4. International Convention for the Prevention of Pollution from Ships (MARPOL) Consolidated
5. International Convention on Load Lines 1996, including Protocol of 1988, including latest amendments
6. International Convention on Tonnage Measurements of Ships
7. International Conference on Revision of the International Regulations for Preventing Collision at Sea (COLREG)
8. International Convention for the Safety of Lives at Sea (SOLAS) of 1974 with Protocols of 1978 and 1988, including latest amendments, where referred by IMO MODU Code

---

1. Contract regulations, rules and standards applicable to automation, control and monitoring are marked in italic text.

9. International Code for the Application of Fire Test Procedures (2010 FTP Code), where required by IMO MODU Code
10. International Code for Fire Safety Systems (FSS Code), where required by IMO MODU Code
11. International Ship and Port Facility Security Code (ISPS)
12. *Relevant regulations of the International Electrical Commission Standards (IEC); IEC 60092 series Standards—Electrical Installations in Ships*
13. *IEC 61892 series Standards—Mobile and fixed offshore units—Electrical Installations*
14. *IEC 60533 Electrical and electronic installations in ships—Electromagnetic compatibility (EMC)*
15. IEC 60079 series Standards—Explosive Atmospheres
16. UK CAA CAP 437 Offshore Helicopter Landing Areas
17. IMO Life Saving Appliances Code (LSA Code) 2003
18. IMO Resolution A468 Noise levels onboard ships
19. USCG PUB 515 "Rules & Regulations for Foreign Vessels Operating in US Waters"
20. International Organization for Standardization ISO 6954 Mechanical vibration—Guidelines for the measurement, reporting and evaluation of vibration with regard to habitability on passenger and merchant vessels
21. USCG NVIC 3-88 for foreign flag MODU
22. International Telecommunication Convention 1973 with Annex and Revisions 1974, 1976, 1979, 1982 and 1983/87 incl. GMDSS
23. UK HSE regulation and recommendation for hull and main equipment
24. UK HSE Guidance Note No. 82 on accommodation
25. UK HSE Offshore Information Sheet No. 1/2006 (Testing Regime for Offshore TR-HVAC Fire Dampers etc.)
26. Global Maritime Distress and Safety System (GMDSS) requirements for sea areas A1+A2+A3
27. UK HSE regulations for pressure vessels
28. The UK Pressure Equipment Regulations 1999 ("the PER"—SI 1999/2001) amended by the Pressure Equipment (Amendment) Regulations 2002 (SI 2002/1267)
29. UK HSE regulations for electrical equipment located in hazardous area
30. European Pressure Equipment Directive ("PED" 97/23/EC)
31. DIRECTIVE 97/23/EC OF THE EUROPEAN PARLIAMENT AND OF THE COUNCIL of 29 May 1997 on the approximation of the laws of the Member States concerning pressure equipment
32. *DIRECTIVE 2014/34/EU OF THE EUROPEAN PARLIAMENT AND OF THE COUNCIL of 26 February 2014 on the harmonization of the laws of the Member States relating to equipment and protective systems intended for use in potentially explosive atmospheres*

33. International Safety Management Code (ISM Code)
34. IMO MSC/Circ. 645 Guidelines for Vessel with Dynamic Positioning System, 1994 (for Class 3 vessels)
35. ABS Publication 86 Guidance Notes for the Application of Ergonomics to Marine Systems

Industrial codes and standards

1. American Institute of Steel Construction (AISC)
2. ISO 7547 Ships and marine technology—Air-conditioning and ventilation of accommodation spaces—Design conditions and basis of calculations
3. ISO 8861 Shipbuilding—Engine-room ventilation in diesel-engined ships—Design requirements and basis of calculations
4. ISO 8862 Air-conditioning and ventilation of machinery control-rooms on board ships—Design conditions and basis of calculations
5. ISO 8864 Air-conditioning and ventilation of wheelhouse on board ships—Design conditions and basis of calculations
6. ISO 9099 Air-conditioning and ventilation of dry provision rooms on board ships—Design conditions and basis of calculations
7. ISO 9943 Shipbuilding—Ventilation and air-treatment of galleys and pantries with cooking appliances
8. ISO 15138 Petroleum and natural gas industries—Offshore production installations—Heating, ventilation and air-conditioning
9. IMO Res. A.468 (XII): Code on noise levels on board ships
10. *DNV-RP-D102; Failure Mode and Effect Analysis (FMEA) of Redundant Systems, January 2012*
11. *DNV-RP-E306; Dynamic Positioning Vessel Design Philosophy Guidelines, September 2012*
12. *IMCA M 103—Guidelines for the Design and Operation of Dynamically Positioned Ships*
13. IMCA M139—Specification for DP Capability Plots
14. IMCA M140—Standard Report for DP Vessel Annual Trials

Appendix 2

Automation DESIGN examples

A2.1 Basic design phase documents examples

A2.1.1 Control and monitoring elements arrangement

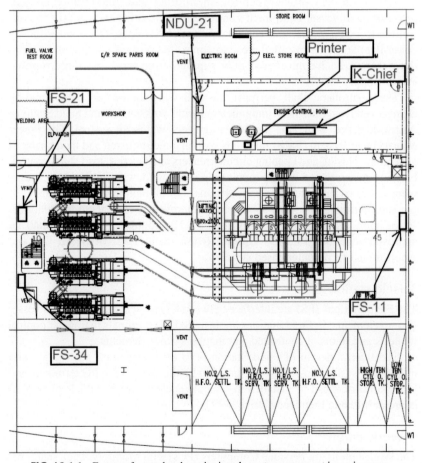

FIG. A2.1.1 Extract of control and monitoring elements arrangement in engine room.

A2.1.2 Automatic vessel control description example

1 Introduction

This Automatic Vessel Control (AVC) description outlines requirements for elements of control system hardware and performed by them functions.

2 Controllers

The controllers are to be marine type working in a duty/stand-by mode. The stand-by controller repeats duty controller actions and continually checks its proper functionality. When duty controller fails than stand-by controller shall act as duty controller and takes care on all system control functions. All controllers shall be connected with Field Stations (FS) via dual Ethernet network.

The controllers provide control, monitoring and alarm functions. All processes and vessel equipment control and monitoring data is sent to controllers from Field Stations via Dual Ethernet Network. Duty and stand-by controllers are to be installed close to controlled process equipment in separate cubicles. Each controller shall be supplied from separate Uninterruptible Power Supply (UPS).

The controllers shall perform the following functions: Duty/Stand-by changeover, Dual Ethernet communication with Workstations and Field Stations, Alarm detection with time tagging, Engineers Watchcall system. The Power Management System (PMS) functions shall include start/stop and monitoring of generators, closing and opening circuit breakers and bus ties, load share control, blackout recovery and engine monitoring with shutdown. The Vessel management functions shall include vessel systems control, monitoring and alarm.

The controllers hardware shall be type approved by member of International Association Classification Society (IACS). The controllers software shall be approved according to IACS Unifier Requirement 'E22 On Board Use and Application of Computer based systems'.

3 Field stations

The Field Stations (FS) are interface between Controllers and controlled vessel system elements. In Field Stations are installed I/O terminals to transfer control requests and status of controlled automation system elements.

The main function of Field Station is to connect control automation and monitoring system sensors and actuators with the controllers. I/O terminals and serial converters can be installed in separate enclosure or in cubicle of switchboard or Variable Frequency Drive (VFD).

4 Workstations

The Workstations are Personal Computers (PCs) with Monitors. They shall provide vessel crew with system/equipment control status and alarm signals using MIMICs on Workstation Monitors.

The Workstations communicate with controllers over the dual Ethernet via Open Platform Communication (OPC) protocol using dual Ethernet driver.

5 Dual Ethernet network

Two separate Ethernet networks shall provide communication between Controllers, Field Stations and Workstations. Networks shall be redundant preserving communication in case of communication failure.

The Dual Communication Network is resistant to multiple faults and on both networks communication is checked several times a second to ensure fast changeover in case of communication failure.

Automatic Vessel Control (AVC) System Dual Ethernet Network transfer rate shall be 100 Mbits/s. Networks communicates AVC controllers with Field Stations, Workstations and may be configured as 'Star' or 'Ring' and may be built using cooper cable or fiber optic connections. Cooper cable runs length is limited to 80 meters between connecting points and Fiber optic connections can be used up to 2000 meters. Additionally, Fiber optic connections shall be used when network crosses A60 barrier, in places with noisy EMC environment or where segregation from power cables is difficult or not possible.

Automatic Vessel Control communication shall be arranged on two separate Ethernet networks run on opposite side of the vessel where is practicable and possible. All network equipment shall be duplicated to ensure communication when one network fails. Both networks shall be connected to each communication node so controllers, Field Stations or Workstations can switch over to other network if one failed. Both networks shall be diagnosed for failures. Both transmit and receive paths shall be checked independently several times a second. Networks shall be protected from being down caused by faulty nodes. Status of both networks shall be shown on Automatic Vessel Control (AVC) Workstation MIMIC's and any failure shall be reported by Alarm Monitoring System. Additionally network statistics shall be available. Controllers, Field Stations and Workstations shall be connected to Dual Ethernet Network by switches enabling flow of data across a network.

A2.1.3 Power management system (PMS) description example

1 Introduction

Power Management System (PMS) shall serve to vessel power station presented on Single line Diagram (SLD) including eight (8) off 2550 kW diesel engines with their 6.6 kV generators, four (4) MV switchboards supplied by two (2) generator sets feeding thrusters and service transformers, eight (8) off fixed speed variable pitch thruster motors fed by their soft-start units, one (1) main propulsion drive driven directly from a diesel engine 1305 kW and LV switchboards supplying auxiliary equipment and loads.

2 Power system operation modes

The vessel shall operate with: DP3 mode with all MV interconnections open using the MV system to power vessel thrusters and LV Switchboards, Normal mode with all MV interconnections closed using the MV system to power vessel thrusters and LV Switchboards and Harbour mode with MV system is shut off and LV power is provided by auxiliary generator or shore supply.

3 Functionality overview

Power Management System (PMS) shall fulfil the following functional requirements:

a. Required MIMICs:

- Generator loading and control providing the MV Power Management interface
- MV switchboards and circuit breaker status
- LV switchboards and circuit breaker status
- Emergency switchboards and circuit breaker status.

b. Required Circuit breakers and bus ties control functions:

- Opening and closing a number of MV switchboard circuit breakers
- Opening and closing a number of principal LV switchboard circuit breakers (for black-out recovery)
- Opening and closing Emergency Switchboard circuit breakers (for black-out recovery)
- Automatic synchronising of the MV interconnector circuit breakers via dedicated auto synchronisers in the MV SWBD (initiated by the PMS)
- Possibility to check interconnector synchroniser in PMS hardware
- Bus section over and under frequency alarms
- Bus section over and under voltage alarms.

c. Required MV Generators control and monitoring functions:

- Initiate starting and stopping of generators
- Initiate automatic generators synchronisation under PMS control. (Automatic synchronising equipment is part of the switchboard supply)
- Engine monitoring (essential parameters e.g. lube oil pressure, cooling water temperature) with alarm generation
- Engine exhaust gas temperature monitoring and deviation checks.

d. Required MV Power management functions:

- Initiate starting and stopping of MV generators depending on load demand. (Automatic synchronising equipment is part of the switchboard supply)
- Control the MV power system steady state voltage to an operator pre-set
- Symmetric sharing of reactive power load between interconnected generating sets

- Asymmetric sharing of reactive power load between interconnected generating sets
- Reactive power load share failure alarm.

e. Required overload protection and load shed functions:

- Initiate load shedding in order to avoid a power system blackout when the system detects busbar under frequency
- Initiate load shedding in order to avoid a power system blackout when the system detects a substantial overload
- Initiate load shedding if the system detects a generator trip
- Provide available load/spinning reserve output signal for each MV bus to allow the DP system to limit load and prevent blackout.

f. Required Blackout recovery functions:

- Restore MV busbar power in the event of a system blackout
- Restore LV power in the event of a system blackout
- Restart essential service pumps and reconnect essential loads upon restoration of supply after a blackout.

g. Required load shed function:

- Non-essential loads shall be shed if Generator trips, Generator is overloaded or is system under frequency.

4 Information presented on MIMICs

The following information shall be presented on MIMICs:

- Individual generator active power
- Individual generator reactive power
- Individual generator frequency
- Individual generator voltage
- Bus section frequency
- Bus section voltage
- Spinning reserve active power
- Spinning reserve reactive power
- Power consumed active power
- Power consumed reactive power
- Minimum number of generators to run
- Generator status, e.g. starting, load sharing, etc.
- Generator start priority
- Generator stop priority
- Generator operating mode
- Generator active power rating
- Generator reactive power rating
- Generator reactive power set point

- Generator active power load
- Generator reactive power load
- Generator active power percentage load
- Generator reactive power percentage load
- Generator current
- Generator power factor
- Cable bus tie current flow.

5 Required control functions

- Start/load up/unload/stop/fast stop generators
- Set minimum number of generators to run
- Enable/disable automatic (load dependent) start of generators
- Enable/disable automatic (load dependent) stop of generators
- Modify generator start/stop priority order
- Set generator to fixed target load mode (kVAr)
- Enable/disable load-shed functions
- Enable/disable blackout recovery

6 Required blackout recovery

The blackout recovery shall recover MV power and restart previously running thrusters.

Emergency generator shall support essential vessel auxiliaries until the LV system is manually reconfigured to supply MV power.

Blackout recovery shall be initiated for total loss of MV generation (complete blackout) or loss of one of the MV switchboards (partial blackout).

A2.1.4 Alarm and monitoring system description example

1 Introduction

The alarm and monitoring system shall be integrated with Automatic Vessel Control AVC system.

The system shall provide the following functions:

- Handling digital and analogue signals
- Configurable by authorized person alarm parameters
- Time tagged alarm signals and events
- Clear alarm message text description including time & date
- Alarm status indication (active/reset, acknowledged, normal)
- Permanently visible alarm bar with alarm statistics
- Available alarm pages for browsing potential alarms, current alarms and logged alarms/events
- Filtering alarms facilities

- – Alarm grouping and sub-grouping
- – Alarm blocking
- – Online or on demand alarm and event printing
- – Engineers watch call alarm system
- – Deadman system.

On Alarm System Monitors shall be visible bars showing active alarms and the total number of active alarms, unacknowledged alarms and acknowledged alarms.

Dedicated alarms pages shall allow the operator to view current alarms, logged alarms or potential alarms from a full sized screen.

2 Alarm groups

The following alarm groups are to be defined on the vessel: Group No 1 Power System—Prime Movers & Gens, Group No 2 Power System—Engines, Group No 3 Power Distribution, Group No 4 Auxiliary System, Group No 5 Fire Detection, Group No 6 Ballast, Group No 7 Thruster System, Group No 8 Propulsion/Main Engine, Group No 9 System Monitoring, Group, No 10 Watertight Doors, Group No 11 HVAC & Cooling and Group No 12 Bilge.

Watch calling system panels shall be installed on the Navigation Bridge, in Duty Engineers No 1 to 4 cabins and seven (7) Public spaces.

3 Audible alarms

Each new alarm shall give audible signalization. Alarm could be muted using dedicated function key on keyboard even if no operator is logged in on Workstation. The mute remains until all alarms are acknowledged or until new alarms is generated.

4 Alarm acknowledgement

Alarm can be acknowledged by logged in operator. Alarm(s) can be acknowledged using the 'Acknowledge Alarm' button which that acknowledges the selected alarm, or using the 'Acknowledge Page' button which acknowledges current displayed alarm page.

When active alarm is acknowledged its status on MIMIC diagram changes to 'Acknowledged'. When 'Acknowledged' alarm parameter return to 'Normal' than alarm is reset on MIMIC diagram.

The system shall be protected against overloading with unacknowledged alarms.

5 Alarms details

Latest alarm shall be displayed on top of alarms page and the following data shall be available when clicked on: System Alarm Number with a unique identity for each alarm in the system, Date in MM/DD/YYYY format, Time in hh:mm:ss format, Alarm group, Alarm description and Alarm Status, for example ACTIVE meaning that alarm is active and has not been acknowledged, RESET meaning that alarm became active, changed back to inactive, and has not been acknowledged, ACKNOWLEDGED meaning that alarm is active and has been acknowledged, NORMAL meaning that alarm has changed to inactive state, and has been acknowledged.

Displayed text shall be distinguished according to the following color scheme: white on red for vital alarms and black on amber for non-vital alarms.

Alarms shall be displayed until they are acknowledged and reset—returned to normal.

6 Alarms visualization

Alarms visualization shall be supported with scroll bar to allow navigation trough the list of alarms. Selected alarm shall be highlighted. Alarms filtering facilities to be provided for selected alarms browsing. The operator may filter alarms on "Priority", "Status", "Group", "Mimic No.", "Time and Date" or by using a text search in the "Contains" or "Does Not Contain" field. When alarms are filtered than on the monitor shall be easy noticeable string with text: 'ALARMS FILTERED'.

7 Alarms printing

Alarms printing shall be initiated by pushing 'Alarms Print Page' button or 'Alarms Print All' button.

8 Logged alarms

Each alarm and event data shall be stored at the workstation hard disk to allow retrieval later. It should be possible to view the logged alarms and events online at any workstation. Additionally it should be possible to print in chronological order logged alarms. Alarm printing color depends on 'Alarm priority' and shall be similar as is on NIMIC diagram. Special filters shall be used to sort and filter alarms to be displayed and/or printed.

Storing at least ten thousand (10 000) alarms and events on the alarm server Workstation's hard disk should be provided.

9 Alarm disabling/blocking

Manual alarm disabling/blocking function shall be provided. This feature may be useful when alarm requirements change and an alarm is required to be removed. It can also be used to enable unused level alarms on an analogue alarm. For analogue signals it should be possible to enable/disable each level alarm individually. Additionally it should be possible to temporarily disable signals from generating any alarm(s) for a time period up to 24 hours. Once the disabling/blocking time expiry the alarm is enabled again. This feature is useful when is carried out automation system maintenance work.

Additionally automatic alarms disabling/blocking shall be provided for alarms that do not need to be detected due to a certain state of controlled or monitored system, for example when a diesel engine is stopped it will usually lose its lube oil pressure and subsequently the lube-oil-pressure-low-alarm will need to be disabled/blocked by the Engine Running signal.

10 Trends monitoring

Analogue and digital signals shall be continuously logged to enable trends analysis.

11 Reports printing

Reports printing facility shall be provided. Reports can be printed periodically or on demand. An operator could select scope of printed report and printing interval time for periodically printed reports.

Periodically printed reports are activated by set 'Periodical time' and on demand printed reports are activated by 'Print' button' and selecting scope of printed report.

12 System health

To determine system health the following system parameters shall be monitored: network alarms, serial link alarms, controller alarms, field station alarms, workstation alarms and internal system alarms.

13 MIMICs colors

Table in Fig. A2.1.2 presents control and monitoring systems animation colors that shall be used on MIMICs to show automation equipment status.

Status of system animation	Color
Running/energized/open	Green
Not running or not energized	White
Fault alarm/closed	Red
Warning alarm	Amber
Blocked	Cyan
Bad data	Blue

FIG. A2.1.2   MIMIC systems animation colors.

Table in Fig. A2.1.3 presents colors assigned to control and monitoring systems media that shall be used on MIMICs.

Media	Color
Steam	White
Potable water medium	Blue
Fresh water medium	Blue
Fresh water (jacket CW)	Medium blue
Sea water	Green
Bilge	Black
Ballast	Dark green
Fire main	Red
Diesel fuel oil	Light brown
Heavy fuel oil	Dark brown
Hydraulic oil light	Brown
Lubricating oil	Yellow
Cargo oil	Brown
Compressed air	Dark violet
Starting air	Cyan
Inert gas	Magenta

FIG. A2.1.3   Colors assigned to systems media on MIMICs.

14 Preliminary list of MIMICs

Preliminary list of MIMICs and their numbers are included in Fig. A2.1.4.

MIMIC no	MIMIC name
No 0	Vessel control menu
No 1	Diesel engine no 1
No 2	Diesel engine no 2
No 3	Diesel engine no 3
No 4	Diesel engine no 4

FIG. A2.1.4   Preliminary list of MIMICs.

No 5	Diesel engine no 5
No 6	Diesel engine no 6
No 7	Diesel engine no 7
No 8	Diesel engine no 8
No 11	Electrical system—Generator 1 & 2
No 12	Electrical system—Generator 3 & 4
No 13	Electrical system—Generator 5 & 6
No 14	Electrical system—Generator 7 & 8
No 20	Electrical system—Generator loading and control
No 21	Electrical system—HV switchboard
No 22	Electrical system—LV switchboard
No 23	Electrical system—Transformer monitoring
No 30	Thruster system—Aft Thruster 1
No 31	Thruster system—Aft Thruster 2
No 32	Thruster system—Azimuth Thruster 3
No 33	Thruster system—Azimuth Thruster 4
No 34	Thruster system—Azimuth Thruster 5
No 35	Thruster system—Azimuth Thruster 6
No 36	Thruster system—Azimuth Thruster 7
No 37	Thruster system—Azimuth Thruster 8
No 38	Thruster system—Azimuth Thruster 9
No 39	Thruster system—Steering gear
No 40	Water system—Ballast system
No 41	Water system—Bilge system
No 42	Water system—fresh water system 1
No 43	Water system—fresh water system 2
No 44	Water system—sea water system 1
No 45	Water system—sea water system 2

FIG. A2.1.4   Cont'd

Continued

No 50	Fuel oil system—Fuel oil service system 1 4
No 51	Fuel oil system—Fuel oil filling and transfer
No 53	Lube oil system—lube oil service system
No 54	Lube oil system—lube oil filling and transfer
No 50	Fire systems—fire systems
No 51	Fire systems—fire zone 1
No 52	Fire systems—fire zone 2
No 53	Watertight doors
No 60	Compressed air—compressed air system
No 70	Auxiliary fans—auxiliary fans 1
No 71	Auxiliary fans—auxiliary fans 2
No 80	Auxiliary pumps—auxiliary pumps 1
No 81	Auxiliary pumps—auxiliary pumps 2
No 90	Running hours—running hours 1-35
No 92	Running hours—running hours 36-70
No 100	System—watchcall
No 101	System—deadman
No 102	System—time zone
No 103	System—system health
No 104	System—control transfer 1
No 105	System—control transfer 2
No 106	System—Control transfer 3
No 107	System—Load shedding 1 9
No 108	System—Server status
No 109	System—Trends
No 110	System—Current alarms
No 111	System—Logged alarms
No 112	System—Potential alarms

FIG. A2.1.4 Cont'd

A2.2 Detail engineering design phase document examples

A2.2.1 Wiring diagrams

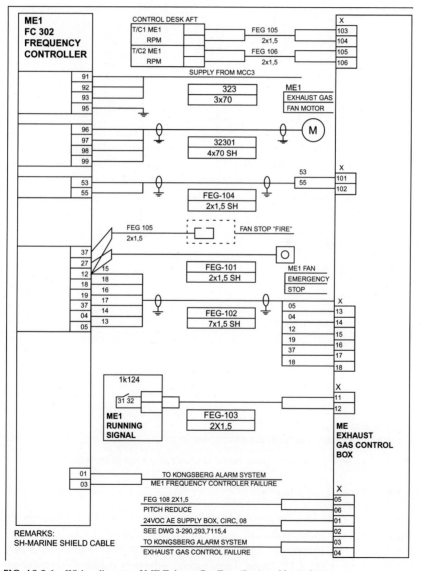

FIG. A2.2.1 Wiring diagram of ME Exhaust Gas Fan. *(Designed by Author.)*

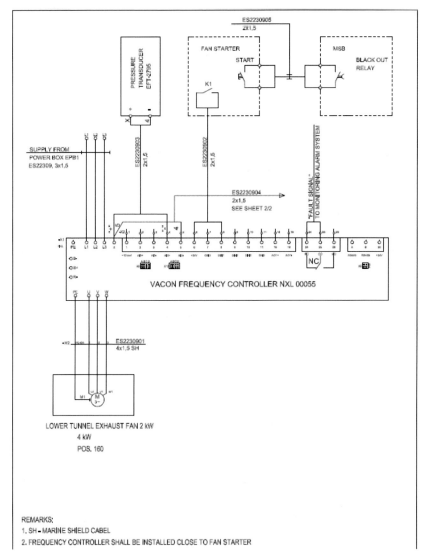

FIG. A2.2.2 Wiring diagram of Lower Tunnel Exhaust Fan. *(Designed by Author.)*

A2.2.2 Termination list

An example of an IAS Termination List is presented in Fig. A2.2.3. The shown Termination List includes the following information: Cable No, Core mark number, Cable type, Instrument Tag, Description, Loop typical, Signal, Normally (closed/open), FTC, FTC Node, Termination, Advantys Address, Termination number and External Unit number.

An example of a typical IAS I/O List is presented in Fig. A2.2.4. The shown I/O list includes the following information: Tag, Signal description, Loop typical, Signal

Cable No	Cabl	Core	Core	Cable Type	Instrument Tag	Description	Log. typ	Sign	Norm	FTC	FTC Node	Termi	Adventys Addre	Terminal	Terminal	iii	External Unit	
875_L15-WI01	part1	1	2	Scr.1x2x0,75mm2	875_L15_351	Distribution Board_L15 Earth Fault Container	DI-301 XA			791.03.02	791.03.02.72	-R72-12	s12.di07 (DDI 3725)	A-14 (24V)	A-15 (0V)		L15 - 230V Distribution Board	
875_L14-WI01	part1		2	Scr.1x2x0,75mm2	875_L14_351	Distribution Board_L14 Earth Fault Container	DI-301 XA		Yes	791.03.03	791.03.03.73	-R73-12	s12.di14 (DDI 3725)	B-12 (24V)	B-13 (0V)		L14 230V Distribution Board	
875-E1-WI01	part1		2	Scr.1x2x0,75mm2	875_010_010_351	Distr. Board E1 loss of main/backup power supply	DI-301 XA		Yes	791.03.01	791.03.01.51	-R51-09	s09.di06 (DDI 3725)	A-12 (24V)	A-13 (0V)		E1 - 230V Emergency Distribution	X22:1
871-IC-WI01	part1/2		2	Scr.2x2x0,75mm2	871_001_005_305	Ambient Air Temp Iron Core Swbd Room Below	AI-303 TIAH		No	791.03.03	791.03.03.53	-R53-08	s08.ai02 (ART 0200)	B-2	B-3	B-5	871_001_005 Temperature Sensor	'I
871-IC.Thr3-WI01	part1/2		2	Scr.2x2x0,75mm2	871_001_004_305	Cable Temp Monitoring Thr. 5 Supply 2	AI-303 TIAH		No	791.03.06	791.03.06.56	-R56-06	s06.ai01 (ART 0200)	A-2	A-3	A-5	871_001_004 Temperature Sensor	'I
871-IC.Thr3-WI01	part1/2		2	Scr.2x2x0,75mm2	871_001_003_305	Cable Temp Monitoring Thr. 4 Supply 1	AI-303 TIAH		No	791.03.06	791.03.06.56	-R56-05	s06.ai02 (ART 0200)	B-2	B-3	B-5	871_001_003 Temperature Sensor	'I
871-IC.Thr4-WI01	part1/2		2	Scr.2x2x0,75mm2	871_001_002_305	Cable Temp Monitoring Thr. 4 Supply 2	AI-303 TIAH		No	791.03.07	791.03.07.57	-R57-06	s06.ai01 (ART 0200)	A-2	A-3	A-5	871_001_002 Temperature Sensor	'I
871-IC.Thr4-WI01	part1/2		2	Scr.2x2x0,75mm2	871_001_001_305	Cable Temp Monitoring Thr. 4 Supply 1	AI-303 TIAH		No	791.03.07	791.03.07.57	-R57-05	s05.ai02 (ART 0200)	B-2	B-3	B-5	871_001_001 Temperature Sensor	'I
867_070.040-WI02	pair1	1		Scr.1x2x0,75mm2	867_070_040_Q40_03_351	Distboard C4 Earth Fault	DI-301 XA		Yes	791.03.02	791.03.02.72	-R72-12	s12.di06 (DDI 3725)	A-12 (24V)	A-13 (0V)		C4 - 230V Distribution Board	
867_070.040-WI01	pair1	1		Scr.1x2x0,75mm2	867_070_040_Q4_351	UPS 4 C4, General Alarm	DI-301 XA		Yes	791.03.02	791.03.02.72	-R72-12	s12.di04 (DDI 3725)	A-7 (24V)	A-8 (0V)		867.070.040 UPS4	-Interfa
867_070.040-WI01	pair2	1		Scr.1x2x0,75mm2	867_070_040_Q2_351	UPS 4 C4, Low Battery Voltage	DI-301 XA		Yes	791.03.02	791.03.02.72	-R72-12	s12.di05 (DDI 3725)	A-10 (24V)	A-11 (0V)		C3 - 230V Distribution Board	-Interfa
867_070.030-WI01	pair1			Scr.2x2x0,75mm2	867_070_030_Q3_351	UPS 3 C3, General Alarm	DI-301 XA		Yes	791.03.03	791.03.03.73	-R73-12	s12.di13 (DDI 3725)	B-10 (24V)	B-11 (0V)		867.070.030 UPS3	-Interfa
867_070.030-WI01	pair2			Scr.2x2x0,75mm2	867_070_030_Q2_351	UPS 3 C3, Low Battery Voltage	DI-301 XA		Yes	791.03.03	791.03.03.73	-R73-12	s12.di11 (DDI 3725)	B-6 (0V)	B-6 (0V)		867.070.030 UPS3	-Interfa
867_070.030-WI01	pair3			Scr.2x2x0,75mm2	867_070_030_Q1_351	Distboard C3 Earth Fault	DI-301 XA		Yes	791.03.03	791.03.03.73	-R73-12	s12.di12 (DDI 3725)	B-7 (24V)	B-8 (0V)		C2 - 230V Distribution Board	
867_070.020-WI02	pair1			Scr.2x2x0,75mm2	867_070_020_04_351	Distboard C2 Earth Fault	DI-301 XA		Yes	791.03.01	791.03.01.51	-R51-09	s09.di04 (DDI 3725)	A-7 (24V)	A-8 (0V)		C2 - 230V Distribution Board	
867_070.020-WI02	pair2			Scr.2x2x0,75mm2	867_070_020_03_351	Distboard C2 Power Failure	DI-301 XA		Yes	791.03.01	791.03.01.51	-R51-09	s09.di05 (DDI 3725)	A-10 (24V)	A-11 (0V)		C2 - 230V Distribution Board	
867_070.020-WI01	pair1			Scr.2x2x0,75mm2	867_070_020_Q2_351	UPS 2 C2, General Alarm	DI-301 XA		Yes	791.03.01	791.03.01.51	-R51-09	s09.di02 (DDI 3725)	A-1 (24V)	A-4 (0V)		867.070.020 UPS2	-Interfa
867_070.020-WI01	pair2			Scr.2x2x0,75mm2	867_070_020_Q1_351	UPS 2 C2, Low Battery Voltage	DI-301 XA		Yes	791.03.01	791.03.01.51	-R51-09	s09.di03 (DDI 3725)	A-5 (24V)	A-6 (0V)		867.070.020 UPS2	-Interfa
867_070.010-WI02	pair1			Scr.2x2x0,75mm2	867_070_010_04_351	Distboard C1 Earth Fault	DI-301 XA		Yes	791.03.01	791.03.01.51	-R51-09	s08.di16 (DDI 3725)	B-16 (24V)	B-17 (0V)		C1 - 230V Distribution Board	
867_070.010-WI02	pair2			Scr.2x2x0,75mm2	867_070_010_03_351	Distboard C1 Power Failure	DI-301 XA		Yes	791.03.01	791.03.01.51	-R51-09	s08.di10 (DDI 3725)	B-1 (24V)	A-2 (0V)		C1 - 230V Distribution Board	
867_070.010-WI01	pair1			Scr.2x2x0,75mm2	867_070_010_Q2_351	UPS 1 C1, General Alarm	DI-301 XA		Yes	791.03.01	791.03.01.51	-R51-08	s08.di14 (DDI 3725)	B-12 (24V)	B-13 (0V)		867.070.010 UPS1	-Interfa
866_B3.010-WI01	pair1			Scr.4x2x0,75mm2	866_B3_010_01_351	UPS 1 C1, Low Battery Voltage	DI-301 XA		Yes	791.03.01	791.03.01.51	-R51-08	s08.di15 (DDI 3725)	B-14 (24V)	B-15 (0V)		867.070.010 UPS1	-Interfa
866_B3.010-WI01	pair1			Scr.4x2x0,75mm2	866_B3_010_04_351	DistroBord B3 Failure Battery Charger No.1 JEG	DI-301 XA		Yes	791.03.01	791.03.01.51	-R51-08	s08.di12 (DDI 3725)	B-7 (24V)	B-8 (0V)		866.B3.010 24VDC B3 Distribution	-X155
866_B3.010-WI01	pair2			Scr.4x2x0,75mm2	866_B3_010_02_351	DistroBord B3 Failure Battery Charger No.2 JEG	DI-301 XA		Yes	791.03.01	791.03.01.51	-R51-08	s08.di13 (DDI 3725)	B-10 (24V)	B-11 (0V)		866.B3.010 24VDC B3 Distribution	
866_B2.010-WI01	pair2			Scr.4x2x0,75mm2	866_B2_010_04_302	DistroBord B2 Failure Battery Charger No.1	DI-301 XA		Yes	791.03.01	791.03.01.51	-R51-08	s08.di10 (DDI 3725)	B-1 (24V)	B-2 (0V)		866.B2.010 24VDC B2 Distribution	
866_B2.010-WI01	pair2			Scr.4x2x0,75mm2	866_B2_010_02_351	DistroBord B2 Failure Battery Charger No.2	DI-301 XA		Yes	791.06.01	791.03.01.51	-R51-08	s08.di10 (DDI 3725)	B-3 (24V)	B-4 (0V)		866.B2.010 24VDC B2 Distribution	
866_B2.010-WI01	pair3			Scr.4x2x0,75mm2	866_B2_010_01_351	DistroBord B2 Insulation Failure	DI-301 XA		Yes	791.03.01	791.03.01.51	-R51-08	s08.di11 (DDI 3725)	B-5 (24V)	B-6 (0V)		866.B2.010 24VDC B2 Distribution	
866_B1.010-WI01	pair1			Scr.4x2x0,75mm2	866_B1_010_04_351	DistroBord B1 Failure Battery Charger No.1	DI-301 XA		Yes	791.03.01	791.03.01.51	-R51-08	s08.di07 (DDI 3725)	A-12 (24V)	A-13 (0V)		866.B1.010 24VDC B1 Distribution	
866_B1.010-WI01	pair2			Scr.4x2x0,75mm2	866_B1_010_02_351	DistroBord B1 Failure Battery Charger No.2	DI-301 XA		Yes	791.03.01	791.03.01.51	-R51-08	s08.di08 (DDI 3725)	A-14 (24V)	A-15 (0V)		866.B1.010 24VDC B1 Distribution	
866_B1.010-WI01	pair3			Scr.4x2x0,75mm2	866_B1_010_01_351	DistroBord B1 Insulation Failure	DI-301 XA		Yes	791.03.01	791.03.01.51	-R51-08	s08.di09 (DDI 3725)	A-16 (24V)	A-17 (0V)		866.B1.010 24VDC B1 Distribution	
865_001.020-CWI01	pair4	7		Scr.(i) 7x2x0,75	865_001_020_02_209	LLC Transformer T2, Fan 2 Running	DI-301 XI		No	791.03.02	791.03.02.72	-R72-11	s11.di15 (DDI 3725)	B-14 (24V)	B-15 (0V)		865.001.020	-X155
865_001.020-CWI01	pair5	10		Scr.(i) 7x2x0,75	865_001_020_02_209	LLC Transformer T2, Fan 2 Running	DI-301 XI		Yes	791.03.02	791.03.02.72	-R72-11	s11.di16 (DDI 3725)	B-16 (24V)	B-17 (0V)		865.001.020	-X155
865_001.020-CWI01	pair5	6		Scr.(i) 7x2x0,75	865_001_004_302	LLC Transformer T2, Water Leakage	DI-301 LAH		Yes	791.03.02	791.03.02.72	-R72-04	s04.ai02 (ACI 1290)	A-1 (24V)	A-2 (0V)		865.001.020	-X148
865_001.020-CWI01	pair1	2		Scr.(i) 7x2x0,75	865_001_020_04_302	LLC Transformer T2, Temp.	AI-302 TIAH/H		No	791.03.02	791.03.02.72	-R72-04	s04.ai02 (ACI 1290)	B-2 (24V)	B-4 (0V)		865.001.020	-X140
865_001.020-CWI01	pair2	4		Scr.(i) 7x2x0,75	865_001_020_05_351	LLC Transformer T2, Temp. Instrument Failure	DI-301 XA		Yes	791.03.02	791.03.02.72	-R72-12	s12.di02 (DDI 3725)	A-3 (24V)	A-4 (0V)		865.001.020	-X142
865_001.010-CWI01	pair6	11		Scr.(i) 7x2x0,75	865_001_020_06_351	LLC Transformer T2, Harmonic Filter Failure	DI-301 XA		Yes	791.03.03	791.03.03.73	-R72-12	s12.di03 (DDI 3725)	A-5 (24V)	A-6 (0V)		865.001.020	-X170
865_001.010-CWI01	pair4	7		Scr.(i) 7x2x0,75	865_001_010_02_209	LLC Transformer T1, Fan 1 Running	DI-301 XI		No	791.03.03	791.03.03.73	-R73-12	s12.di06 (DDI 3725)	A-12 (24V)	A-13 (0V)		865.001.010	-X155

FIG. A2.2.3 Example of IAS Termination list. *(Based on Wartsila documentation.)*

Tag	Signal Description	Loop Typical	Signal Typ	Revision_Comment	Revision No.	Normally Close	Field Termination Cabinet	Eng Unit	Eng Range Low	Eng Range High	LL Limit	L Limit	H Limit	HH Limit	Alarm Grou	Alarm Delay	Command Group
269_011_000_351	LAL Glycol Expansion Tank	DI-3011	LAL			Yes	791.03.05								10	5	12
279_001_351	XOP System Alarm	DI-3011	XA			Yes	791.03.05								14	5	5
288_001_351	Electrolytic Antifouling For Seachests Alarm	DI-3011	XA			Yes	791.03.02								14		5
288_002_351	ICAF System Alarm	DI-3011	XA			Yes	791.03.03								14	5	5
336_001_000_01_360	DB Compressor No. 1 Automatic Operation	DI-3011	XI	Alarm group changed from 14 to 9 BIS	8	No	791.03.05								9		15
336_001_000_02_360	DB Compressor No. 1 Load/Unload	DO-3011	XC	Alarm group changed from 14 to 9 BIS	8	No	791.03.05								9		15
336_001_000_03_360	DB Compressor No. 1 or 2 Run	DO-3011	XC	Alarm group changed from 14 to 9 BIS	8	No	791.03.01								9		15
336_001_000_04_360	DB Compressor No. 1 Remote Press. Sensing	DO-3011	XC	Alarm group changed from 14 to 9 BIS	8	No	791.03.05								9		15
336_001_000_05_360	DB Compressor No. 1 Remote Pressure Selection	DO-3011	XC	Alarm group changed from 14 to 9 BIS	8	No	791.03.05								9		15
336_001_000_203	DB Compressor No. 1 Remote Control	DI-3011	XI	Alarm group changed from 14 to 9 BIS	8	Yes	791.03.05								9		15
336_001_000_207	DB Compressor No. 1 Start/Stop	DO-3011	XC	Alarm group changed from 14 to 9 BIS	8	No	791.03.05								9		15
336_001_000_209	DB Compressor No. 1 Remote Control	DI-3011	XI	Alarm group changed from 14 to 9 BIS	8	No	791.03.05								9	5	15
336_001_000_209	DB Compressor No. 1 Running	DI-3011	XI	Alarm group changed from 14 to 9 BIS	8	No	791.03.05								9		15
336_001_000_219	DB Compressor No. 1 Load/Unload	DI-3011	XI	Alarm group changed from 14 to 9 BIS	8	No	791.03.05								9	5	15
336_001_000_315	Pressure Transmitter DB Compressor No. 1	AI-3021	PiAH	Alarm group changed from 14 to 9 BIS	8	No	791.03.05	bar	0	10			10		9		15
336_001_000_317	DB Compressor No. 1 Current	AI-3021	XA	Alarm group changed from 14 to 9 BIS	8	No	791.03.05	A	0	255					9	5	15
336_001_000_358	DB Compressor No. 1 General Warning	DI-3011	XA	Alarm group changed from 14 to 9 BIS	8	Yes	791.03.05								9	5	15
336_001_000_359	DB Compressor No. 1 General Shutdown	DI-3011	XA	Alarm group changed from 14 to 9 BIS	8	Yes	791.03.05								9		15
336_001_000_02_360	DB Compressor No. 2 Automatic Operation	DI-3011	XI	Alarm group changed from 14 to 9 BIS	8	No	791.03.04								9		15
336_001_000_02_360	DB Compressor No. 2 Load/Unload	DO-3011	XC	Alarm group changed from 14 to 9 BIS	8	No	791.03.04								9		15
336_001_000_04_360	DB Compressor No. 2 Remote Press. Sensing	DO-3011	XC	Alarm group changed from 14 to 9 BIS	8	No	791.03.04								9		15
336_001_000_203	DB Compressor No. 2 Remote Pressure Selection	DO-3011	XC	Alarm group changed from 14 to 9 BIS	8	No	791.03.04								9		15
336_001_000_207	DB Compressor No. 2 Start/Stop	DO-3011	XI	Alarm group changed from 14 to 9 BIS	8	Yes	791.03.04								9		15
336_001_000_209	DB Compressor No. 2 Remote Control	DI-3011	XI	Alarm group changed from 14 to 9 BIS	8	No	791.03.04								9		15
336_001_000_219	DB Compressor No. 2 Running	DI-3011	XI	Alarm group changed from 14 to 9 BIS	8	No	791.03.04								9		15
336_001_000_315	Pressure Transmitter DB Compressor No. 2	AI-302	PiAH	Alarm group changed from 14 to 9 BIS	8	No	791.03.04	bar	0	10			10		9		15
336_001_000_317	DB Compressor No. 2 Current	AI-302	XI	Alarm group changed from 14 to 9 BIS	8	No	791.03.04	A	0	255					9		15
336_001_000_358	DB Compressor No. 2 General Warning	DI-3011	XA	Alarm group changed from 14 to 9 BIS	8	Yes	791.03.04								9		15
336_001_000_359	DB Compressor No. 2 General Shutdown	DO-3011	XA	Alarm group changed from 14 to 9 BIS	8	No	791.03.05								9		15
336_006_000_207	Air Dryer No. 1 Dry Bulk System Start/Stop	DI-SWI01	XI	New Software Tag for Generation BIS	8	No	791.03.05								99		15
336_006_000_207	Air Dryer No. 1 Dry Bulk System Dryer Remote SW Tag	DI-3011	XI	Alarm group changed from 14 to 9 BIS	8	No	791.03.05								9		15
336_006_000_209	Air Dryer No. 1 Dry Bulk System Dryer Running	DI-3011	XA	Alarm group changed from 14 to 9 BIS	8	Yes	791.03.05								9		15
336_006_000_351	Air Dryer No. 1 Dry Bulk System Warning Alarm	DO-3011	XA	Alarm group changed from 14 to 9 BIS	8	Yes	791.03.05								9	5	15
336_006_000_358	Air Dryer No. 1 Dry Bulk System Shutdown Alarm	DO-3011	XA	Alarm group changed from 14 to 9 BIS	8	No	791.03.04								9		15
336_006_000_207	Air Dryer No. 1 Dry Bulk System Dryer Remote SW Tag	DI-SWI01	XI	New Software Tag for Generation BIS	8	Yes	791.03.04								99		15
336_006_000_209	Air Dryer No. 1 Dry Bulk System Dryer Running	DI-3011	XI	Alarm group changed from 14 to 9 BIS	8	No	791.03.04								9	5	15
336_006_000_351	Air Dryer No. 1 Dry Bulk System Warning Alarm	DI-3011	XA	Alarm group changed from 14 to 9 BIS	8	Yes	791.03.04								9		15
336_006_000_358	Air Dryer No. 1 Dry Bulk System Shutdown Alarm	DI-3011	XA	Alarm group changed from 14 to 9 BIS	8	No	791.03.04	Pa	0	10					9	5	15
336_006_000_315	Dust Collector No. 1 Dry Bulk Side	AI-301	PiAH	Alarm group changed from 14 to 9 BIS	8	No	791.03.05										

FIG. A2.2.4 Example of IAS I/O list. (*Based on Wartsila documentation.*)

type, Revision comment, Revision No., Normally Closed/Open, Field Termination Cabinet, Eng. Units, Eng. Range Low, Eng. Range High, LL Limit, L Limit, H Limit, HH Limit, Alarm Group, Alarm Delay, Command Group, Ex Signal, Xref, IAS Page, Serial Line Name, Serial Line Address, Serial line Range Low, Serial Line Range High, Serial line protocol, Serial line transmission, WSD Remarks, Supplier, IO Counted, DWG Supplier and Ex Barrier.

Presented below list of cable indexes ensures a consistent approach to control and monitoring system cables numbering throughout the detail design stage of vessel automation systems.

Control and Monitoring cables shall be numbered using cable indexes shown below and shall be followed by schematic diagram cable sequential number.

Cable index	System
ACO2	Alarm CO2
AE	Alarm engineers call
AFA	Fire alarm
AGA	General alarm
AH	Hospital alarm
AM	Machinery alarm
BC	Bridge control and supervision
CA	Auxiliary machinery control and supervision
CB	Boilers control and supervision
CEG	Emergency generator control and supervision
CER	Engine room control and supervision
CFD	Fire doors control and supervision
CG	Gangway control and supervision
CI	Incinerator control and supervision
CLG	Low voltage generators control and supervision
CMG	Medium voltage generators control and supervision
CP	Purifiers control and supervision
CPM	Power management system control
CR	Cranes control and supervision
CS	Sewage control and supervision
CT	Tanks level control and monitoring
CW	Winches control and supervision
CWD	Watertight doors control and supervision
DP	Dynamic positioning control and supervision
ESD	Emergency shut down system
FD	Fire detection system
GD	Gas detection system
HV	Heating, ventilation, air conditioning
PA	Public address system
RS	Refrigeration system control and supervision
TA	Automatic telephone system
TB	Talk back system
TCC	Closed circuit television system
TS	Sound powered telephones system

FIG. A2.2.5 Example of IAS Control and Monitoring Cables Numbering Principles.

A2.3 Technological documentation examples

A2.3.1 Instruction for control and monitoring cables installation

1 Introduction

The Instruction describes control and monitoring circuit cables installation and segregation to ensure a consistent approach of their routing thought the vessel. Cables shall be segregated to minimize electromagnetic interference between different systems cabling and assure installation electromagnetic compatibility. For the purpose of this instruction cables are divided for the following categories.

2 Communication circuit cables—Category 'C'

C.1 Primary, secondary communication and signal circuit cables to be run on separate dedicated cableway

C.2 General communication circuit cables, for example telephone cables to be run on separate dedicated cableway where practicable. Cables may run on main cable way but must be banded separately and separated at least 100 mm from other cables.

3 Instrumentation circuit cables—Category 'I'

I.1 Instrumentation cables for power circuits 0–48 V d.c. preferable shall be run on an instrumentation cableway or may be run on main cableway separated at least 100 mm from main power cables.

I.2 Digital or Analogue Instrumentation circuit cables preferable shall be run on instrumentation cableway and may be grouped together but shall be banded separately from other Instrumentation cables. These cables may run on main cableway but should be separated at least 100 mm from main power cables.

I.3 Analogue cables with small signal, for example for Thermocouples, RTDs and cables for serial data links shall be run on instrumentation cableway and may be grouped together but shall be banded separately from other Instrumentation cables. These cables may run on main cableway but should be separated at least 150 mm from main power cables.

I.4 Intrinsically Safe Circuit cables preferable shall be run on separate dedicated cableway. These cables may run on instrumentation cableway but must be separated at least 50 mm from other instrumentation cables (preferably 100 mm).

I.5 Fibre optic cables shall be run on main cableway, there is no segregation required but these cables should be visible thought their run for inspection or testing purpose.

I.6 Fire detection cables shall be run on dedicated cableway wherever is practicable. These cables may be run on main cableway but must be separated at least 50 mm from other cables.

4 Power cables for automation systems supply—Category 'P'

P.1 Power cables for 24 V d.c. shall be run on main cableway but should be banded separately from cables over 24 V.

P.2 Control cables 220/115 V shall be run on the main cableway but should be banded separately from main power cables.

A2.3.2 Example for work instruction structured network—Test and termination

1 Safety

Special care to be taken when this work is close to 'live 'cables, is curried in confined space or height level.

Make sure that necessary Work permits' are correct and up-to-date.

Using appropriate document check that the facility is not energized and when apply padlock on the circuit breakers are on place before starting the work.

All works shall be done according to applicable shipyard General Safety Rules.

2 Scope

Work scope includes cables:

- Connection and termination in accordance with termination drawings and equipment manuals
- Testing
- Preparing Quality records

3 Cables testing instruction

Intention of this instruction is to ensure correct and safe testing of cable installation. Supervisor/Foreman is responsible that installer knows the procedure and follows it during work planning and execution. All works including safety measures shall be planned to secure that all tests are performed properly.

During tests the following equipment shall be used:

- Voltage tester.
- Insulation tester.
- Insertion Loss Test Kit—Power Meter OPM5 and Light Source OLS4
- OTDR optical time-domain reflectometer
- Fluke DTX-1800 cable analyzer

3.1 Automation and monitoring power supply cables

All automation and monitoring power supply cables insulation resistance (so called megger test) shall be measured applying D.C. voltage according to DNV GL Rules as follows:

- For cable rated voltage $U_n \leq 250$ V minimum test voltage shall be 2 x U_n and minimum insulation resistance have to be 1 MΩ.
- Testing to be performed against earth and phase cores.

After insulation resistance measurement the testing cables must be short-circuited and discharged to earth.

3.2 Instrumentation and signal cables

All Instrumentation and Signal cables insulation resistance shall be measured using Ω Meter applying D.C. voltage according to DNVGL Rules as follows:

For Instrumentation and Signal cable minimum test voltage shall be 2 x U_n and minimum insulation resistance have to be 1 MΩ.

Additionally continuity of the loops shall be tested.

3.3 Fiber optical cables

Fiber optical cables shall be tested using Insertion Loss Test Kit—Power Meter OPM5 and Light Source OLS4. Fiber optical cable connectors shall be cleaned before testing. For multimode cables shall be used wavelength 850 nm and 1300 nm and for single mode cables shall be used wavelength 1300 nm and 1550 nm. Optical Time-Domain Reflectometer (OTDR) shall be used for recording actual length of tested cable and shall be input to Test Result Manager (TRM).

Test reports shall be generated by Test Result Manager (TRM) and shall include pass/fail limits according to relevant applications.

3.4 Network cables CATx type

Copper network cables are to be tested using Fluke DTX1800 Cable Analyzer.

Test parameters shall be set according control and monitoring system connections required specification. Test results shall be generated by DTX Cable Analyzer and shall include pass/fail limits according to relevant applications.

3.5 Coax cables

Coax cables shall be tested for continuity of center and screen after their termination to control and monitoring system elements.

4 List of cables to be tested

List of cables to be tested is presented in Fig. A2.3.1.

Cable number	From cabinet	Patch panel	To
425PA201-T01	425-RQ-003A	425-RQ-031B/1	425-RP-301
425PA202-T01	425-RQ-003A	425-RQ-031B/2	425-RP-302
425PA203-T01	425-RQ-003A	425-RQ-031B/3	425-RP-303
425PA204-T01	425-RQ-003A	425-RQ-031B/4	425-RP-304

FIG. A2.3.1   List of cables to be tested.

5 Installation check list

Installation check list is presented in Fig. A2.3.2.

Installation	Carried out [Yes/not] Y/N	Not checked [Yes/not] Y/N	Not applicable [Yes/not] Y/N
MCT/glands			
Terminating			
Testing			
Cable marking			
Equipment marking			
Cleaning			
Installer signature and date			
Foreman signature and date			
Notes:			

FIG. A2.3.2   Installation check list.

6 Quality records—cable test results

Cable test results are presented in Fig. A2.3.3.

Cable number	Cable type	Passed/failed [P/F]	Comment
425PA201-T01	Cat C7 LSHF-FR 4x2/0.27 mm^2		
425PA202-T01	Cat C7 LSHF-FR 4x2/0.27 mm^2		
425PA203-T01	Cat C7 LSHF-FR 4x2/0.27 mm^2		
425PA204-T01	Cat C7 LSHF-FR 4x2/0.27 mm^2		
Tester signature and date			
Supervisor signature and date			

FIG. A2.3.3 Cable test results.

Appendix 3

Web references

Chapter 2

1. API webpage http://www.api.org/
2. https://www.standard.no/en/sectors/energi-og-klima/petroleum/norsok-standards/#.W89PB2dDtis

Chapter 5

1. IACS web page http://www.iacs.org.uk/
2. Based on ABS web page https://ww2.eagle.org/en/rules-and-resources/rules-and-guides.html#/content/dam/eagle/rules-and-guides/current/conventional_ocean_service/2_steel_vessel_rules_2018
3. Based on ABS web page https://ww2.eagle.org/en/rules-and-resources/rules-and-guides.html#/content/dam/eagle/rules-and-guides/current/offshore/6_mobile_offshore_drilling_units_2018
4. Based on DNVGL web page https://rules.dnvgl.com/ServiceDocuments/dnvgl/#!/industry/1/Maritime/1/DNV%20GL%20rules%20for%20classification:%20Ships%20(RU-SHIP)
5. Based on DNVGL web page https://rules.dnvgl.com/ServiceDocuments/dnvgl/#!/industry/2/Oil%20and%20Gas/14/DNV%20GL%20standards%20(ST)
6. Based on DNVGL web page https://rules.dnvgl.com/ServiceDocuments/dnvgl/#!/industry/2/Oil%20and%20Gas/14/DNV%20GL%20standards%20(ST)

Chapter 6

1. API webpage http://www.api.org/
2. IEC web page http://www.iec.ch/standardsdev/publications/is.htm
3. IEC Webstore web page https://webstore.iec.ch/?ref=menu

4. Based on webpage https://www.ieee.org/index.html
5. Based on webpage http://eur-lex.europa.eu/homepage.html
6. Based on webpage http://www.api.org/
7. Based on webpage https://www.standard.no/en/sectors/energi-og-klima/ petroleum/norsok-standards/#.WpgHzmepXYU

Chapter 7

1. Based on Petroleum Safety Authority YA-711 Principles for alarm system design www.ptil.no/getfile.php/135975/
2. Norwegian Oil and Gas Association Guideline 070 https://www.norskoljeoggass. no/en/working-conditions/retningslinjer/
3. Kongsberg webpage https://www.km.kongsberg.com/ks/web/nokbg0397. nsf/AllWeb/007C8A66DC2782F7C1257B9600247725/$file/350840a. pdf?OpenElement

Chapter 9

1. Kongsberg webpage https://www.km.kongsberg.com/ks/web/nokbg0397. nsf/AllWeb/007C8A66DC2782F7C1257B9600247725/$file/350840a. pdf?OpenElement

Chapter 11

1. Brochure on Kongsberg webpage https://www.km.kongsberg.com/ks/web/ nokbg0240.nsf/AllWeb/B34BE0784AD54E48C12572660040AAFA?Open Document
2. Based on Wartsila webpage https://cdn.wartsila.com/docs/default-source/ product-files/ea/automation/brochure-o-ea-valmatic.pdf?utm_source= autnavdp&utm_medium=automation&utm_term=valmaticplatinum&utm_ content=brochure&utm_campaign=msleadscoring

Chapter 12

1. Brochure from Jotron Webpage https://www.jotron.com/Artikkel/Brochures/ Maritime-&-Energy/10002032.php
2. Based on Jotron Marine & Energy Division Product Catalogue http://pdf. nauticexpo.com/pdf/jotron/product-catalogue-july-2017/22206-100695. html

Chapter 13

1. Kongsberg webpage for K-Pos DP Product Description https://www.km.kongsberg.com/ks/web/nokbg0240.nsf/AllWeb/14E17775E088ADC2C1256A4700319B04?OpenDocument
2. From Kongsberg webpage https://www.km.kongsberg.com/ks/web/nokbg0240.nsf/AllWeb/0A0C3F74B421A7DEC1256A49002DA456?OpenDocument

Chapter 14

1. Based on Kongsberg webpage https://www.km.kongsberg.com/ks/web/nokbg0240.nsf/AllWeb/F5D0D79A0023B87CC12579EA0043A744?OpenDocument
2. Based on ABB webpage https://library.e.abb.com/public/97e02350b7e6330bc1257c47004b1622/RDS%20Marine_Brochure%202014.pdf
3. Based on ABB brochure http://www04.abb.com/global/seitp/seitp202.nsf/0/4b29c4cc981a6e1ec1257bfe00372f5c/$file/27+Sep+2013_ABB+Marine+Remote+Diagnostic+Services.pdf

Chapter 15

1. Based on webpage http://www.iacs.org.uk/

Chapter 16

1. IEC web page http://www.iec.ch/standardsdev/publications/is.htm

Chapter 17

1. Based on webpage https://www.standard.no/en/sectors/energi-og-klima/petroleum/norsok-standards/#.WpgHzmepXYU

Appendix 4

International Electro-technical Committee (IEC) Standards referenced in ship and mobile offshore Unit automation systems[1]

IEC 60092-101 Electrical installations in ships—Part 101: Definitions and general requirements

IEC 60092-201 Electrical installations in ships—Part 201: System design—General

IEC 60092-202 Electrical installations in ships—Part 202: System design—Protection

IEC 60092-301 Electrical installations in ships—Part 301: Equipment—Generators and motors

IEC 60092-302 Electrical installations in ships—Part 302: Low-voltage switchgear and controlgear assemblies

IEC 60092-303 Electrical installations in ships—Part 303: Equipment—Transformers for power and

IEC 60092-304 Electrical installations in ships—Part 304: Equipment—Semiconductor convertors

IEC 60092-305 Electrical installations in ships—Part 305: Equipment—Accumulator (storage) batteries

IEC 60092-350 Electrical installations in ships—Part 350: General construction and test methods of power, control and instrumentation cables for shipboard and offshore applications

IEC 60092-352 Electrical installations in ships—Part 352: Choice and installation of electrical cables

---

1. Latest revision of IEC standards shall be used.

IEC 60092-353 Electrical installations in ships—Part 353: Power cables for rated voltages 1 kV and 3 kV

IEC 60092-360 Electrical installations in ships—Part 360: Insulating and sheathing materials for shipboard and offshore units, power, control, instrumentation and telecommunication cables

IEC TR 60092-370 Electrical installations in ships—Part 370: Guidance on the selection of cables for telecommunication and data transfer including radio-frequency cables

IEC 60092-376 Electrical installations in ships—Part 376: Cables for control and instrumentation circuits 150/250 V (300 V)

IEC 60092-401 Electrical installations in ships—Part 401: Installation and test of completed installation

IEC 60092-501 Electrical installations in ships—Part 501: Special features—Electric propulsion plant

IEC 60092-502 Electrical installations in ships—Part 502: Tankers—Special features

IEC 60092-504 Electrical installations in ships—Part 504: Automation, control and instrumentation

IEC 60092-506 Electrical installations in ships—Part 506: Special features—Ships carrying specific dangerous goods and materials hazardous only in bulk

IEC 60092-509 Electrical installations in ships—Part 509: Operation of electrical installations

IEC 61892-3 Mobile and fixed offshore units—Electrical installations—Part 3: Equipment

IEC TR 62482 Electrical installations in ships—Electromagnetic compatibility—Optimising of cable installations on ships—Testing method of routing distance

IEC 61892-1 Mobile and fixed offshore units—Electrical installations—Part 1: General requirements and conditions

IEC 61892-2 Mobile and fixed offshore units—Electrical installations—Part 2: System design

IEC 61892-3 Mobile and fixed offshore units—Electrical installations—Part 3: Equipment

IEC 61892-4 Mobile and fixed offshore units—Electrical installations—Part 4: Cables

IEC 61892-5 Mobile and fixed offshore units—Electrical installations—Part 5: Mobile units

IEC 61892-6 Mobile and fixed offshore units—Electrical installations—Part 6: Installation

IEC 61892-7 Mobile and fixed offshore units—Electrical installations—Part 7: Hazardous areas

IEC 60533 Electrical and electronic installations in ships—Electromagnetic compatibility (EMC)—Ships with a metallic hull

IEC 60079-0 Explosive atmospheres—Part 0: Equipment—General requirements

IEC 60079-1:2014 Explosive atmospheres—Part 1: Equipment protection by flameproof enclosures "d"

IEC 60079-2 Explosive atmospheres—Part 2: Equipment protection by pressurized enclosure "p"

IEC 60079-5 Explosive atmospheres—Part 5: Equipment protection by powder filling "q"

IEC 60079-6 Explosive atmospheres—Part 6: Equipment protection by liquid immersion "o"

IEC 60079-7 Explosive atmospheres—Part 7: Equipment protection by increased safety "e"

IEC 60079-10-1 Explosive atmospheres—Part 10-1: Classification of areas—Explosive gas atmospheres

IEC 60079-10-2 Explosive atmospheres—Part 10-2: Classification of areas—Explosive dust atmospheres

IEC 60079-11 Explosive atmospheres—Part 11: Equipment protection by intrinsic safety "i"

IEC 60079-13 Explosive atmospheres—Part 13: Equipment protection by pressurized room "p" and artificially ventilated room "v"

IEC 60079-14 Explosive atmospheres—Part 14: Electrical installations design, selection and erection

IEC 60079-15 Explosive atmospheres—Part 15: Equipment protection by type of protection "n"

IEC TR 60079-16 Electrical apparatus for explosive gas atmospheres. Part 16: Artificial ventilation for the protection of analyser(s) houses

IEC 60079-17 Explosive atmospheres—Part 17: Electrical installations inspection and maintenance

IEC 60079-18 Explosive atmospheres—Part 18: Equipment protection by encapsulation "m"

IEC 60079-19 Explosive atmospheres—Part 19: Equipment repair, overhaul and reclamation

IEC 60079-25 Explosive atmospheres—Part 25: Intrinsically safe electrical systems

IEC 60079-26 Explosive atmospheres—Part 26: Equipment with Equipment Protection Level (EPL) Ga

IEC 60079-28 Explosive atmospheres—Part 28: Protection of equipment and transmission systems using optical radiation

IEC 60079-29-1 Explosive atmospheres—Part 29-1: Gas detectors—Performance requirements of detectors for flammable gases

IEC 60079-29-2 Explosive atmospheres—Part 29-2: Gas detectors—Selection, installation, use and maintenance of detectors for flammable gases and oxygen

IEC 60079-29-3 Explosive atmospheres—Part 29-3: Gas detectors—Guidance on functional safety of fixed gas detection systems
IEC 60079-29-4 Explosive atmospheres—Part 29-4: Gas detectors—Performance requirements of open path detectors for flammable gases
IEC 60079-31 Explosive atmospheres—Part 31: Equipment dust ignition protection by enclosure "t"
IEC TS 60079-32-1 Explosive atmospheres—Part 32-1: Electrostatic hazards, guidance
IEC 60079-32-2 Explosive atmospheres—Part 32-2: Electrostatics hazards—Tests
IEC 60079-33 Explosive atmospheres—Part 33: Equipment protection by special protection 's'
IEC TS 60079-39 Explosive atmospheres—Part 39: Intrinsically safe systems with electronically controlled spark duration limitation
IEC TS 60079-40 Explosive atmospheres—Part 40: Requirements for process sealing between flammable process fluids and electrical systems
IEC TS 60079-43 Explosive atmospheres—Part 43: Equipment in adverse service conditions
IEC TS 60079-46 Explosive atmospheres—Part 46: Equipment assemblies
IEC TR 61000-4-1 Electromagnetic compatibility (EMC)—Part 4-1: Testing and measurement techniques—Overview of IEC 61000-4 series
IEC 61000-4-2 Electromagnetic compatibility (EMC)—Part 4-2: Testing and measurement techniques—Electrostatic discharge immunity test
IEC 61000-4-3 Electromagnetic compatibility (EMC)—Part 4-3: Testing and measurement techniques—Radiated, radio-frequency, electromagnetic field immunity test
IEC 61000-4-4 Electromagnetic compatibility (EMC)—Part 4-4: Testing and measurement techniques—Electrical fast transient/burst immunity test
IEC 61000-4-5 Electromagnetic compatibility (EMC)—Part 4-5: Testing and measurement techniques—Surge immunity test
IEC 61000-4-6 Electromagnetic compatibility (EMC)—Part 4-6: Testing and measurement techniques—Immunity to conducted disturbances, induced by radio-frequency fields
IEC 61000-4-7 Electromagnetic compatibility (EMC)—Part 4-7: Testing and measurement techniques—General guide on harmonics and interharmonics measurements and instrumentation, for power supply systems and equipment connected thereto
IEC 61000-4-8 Electromagnetic compatibility (EMC)—Part 4-8: Testing and measurement techniques—Power frequency magnetic field immunity test
IEC 61000-4-9 Electromagnetic compatibility (EMC)—Part 4-9: Testing and measurement techniques—Impulse magnetic field immunity test
IEC 61000-4-10 Electromagnetic compatibility (EMC)—Part 4-10: Testing and measurement techniques—Damped oscillatory magnetic field immunity test

IEC 61000-4-11 Electromagnetic compatibility (EMC)—Part 4-11: Testing and measurement techniques—Voltage dips, short interruptions and voltage variations immunity tests

IEC 61000-4-12 Electromagnetic Compatibility (EMC)—Part 4-12: Testing and measurement techniques—Ring wave immunity test

IEC 61000-4-13 Electromagnetic compatibility (EMC)—Part 4-13: Testing and measurement techniques—Harmonics and interharmonics including mains signaling at a.c. power port, low frequency immunity tests

IEC 61000-4-14 Electromagnetic compatibility (EMC)—Part 4-14: Testing and measurement techniques—Voltage fluctuation immunity test for equipment with input current not exceeding 16 A per phase

IEC 61000-4-15 Electromagnetic compatibility (EMC)—Part 4-15: Testing and measurement techniques—Flickermeter—Functional and design specifications

IEC 61000-4-16 Electromagnetic compatibility (EMC)—Part 4-16: Testing and measurement techniques—Test for immunity to conducted, common mode disturbances in the frequency range 0 Hz to 150 kHz

IEC 61000-4-17 Electromagnetic compatibility (EMC)—Part 4-17: Testing and measurement techniques—Ripple on d.c. input power port immunity test

IEC 61000-4-18 Electromagnetic compatibility (EMC)—Part 4-18: Testing and measurement techniques—Damped oscillatory wave immunity test

IEC 61000-4-19 Electromagnetic compatibility (EMC)—Part 4-19: Testing and measurement techniques—Test for immunity to conducted, differential mode disturbances and signalling in the frequency range 2 kHz to 150 kHz at a.c. power ports

IEC 61000-4-20 Electromagnetic compatibility (EMC)—Part 4-20: Testing and measurement techniques—Emission and immunity testing in transverse electromagnetic (TEM) waveguides

IEC 61000-4-21 Electromagnetic compatibility (EMC)—Part 4-21: Testing and measurement techniques—Reverberation chamber test methods

IEC 61000-4-22 Electromagnetic compatibility (EMC)—Part 4-22: Testing and measurement techniques—Radiated emissions and immunity measurements in fully anechoic rooms (FARs)

IEC 61000-4-2 Electromagnetic compatibility (EMC)—Part 4-23: Testing and measurement techniques—Test methods for protective devices for HEMP and other radiated disturbances

IEC 61000-4-24 Electromagnetic compatibility (EMC)—Part 4-24: Testing and measurement techniques—Test methods for protective devices for HEMP conducted disturbance

IEC 61000-4-25 Electromagnetic compatibility (EMC)—Part 4-25: Testing and measurement techniques—HEMP immunity test methods for equipment and systems

IEC 61000-4-27 Electromagnetic compatibility (EMC)—Part 4-27: Testing and measurement techniques—Unbalance, immunity test for equipment with input current not exceeding 16 A per phase

IEC 61000-4-28 Electromagnetic compatibility (EMC)—Part 4-28: Testing and measurement techniques—Variation of power frequency, immunity test for equipment with input current not exceeding 16 A per phase

IEC 61000-4-29 Electromagnetic compatibility (EMC)—Part 4-29: Testing and measurement techniques—Voltage dips, short interruptions and voltage variations on d.c. input power port immunity tests

IEC 61000-4-30 Electromagnetic compatibility (EMC)—Part 4-30: Testing and measurement techniques—Power quality measurement methods

IEC 61000-4-31 Electromagnetic compatibility (EMC)—Part 4-31: Testing and measurement techniques—AC mains ports broadband conducted disturbance immunity test

IEC TR 61000-4-32 Electromagnetic compatibility (EMC)—Part 4-32: Testing and measurement techniques—High-altitude electromagnetic pulse (HEMP) simulator compendium

IEC 61000-4-33 Electromagnetic compatibility (EMC)—Part 4-33: Testing and measurement techniques—Measurement methods for high-power transient parameters

IEC 61000-4-34 Electromagnetic compatibility (EMC)—Part 4-34: Testing and measurement techniques—Voltage dips, short interruptions and voltage variations immunity tests for equipment with mains current more than 16 A per phase

IEC TR 61000-4-35 Electromagnetic compatibility (EMC)—Part 4-35: Testing and measurement techniques—HPEM simulator compendium

IEC 61000-4-36 Electromagnetic compatibility (EMC)–Part 4-36: Testing and measurement techniques—IEMI immunity test methods for equipment and systems

IEC TR 61000-4-37 Electromagnetic compatibility (EMC)—Calibration and verification protocol for harmonic emission compliance test systems

IEC TR 61000-4-38 Electromagnetic compatibility (EMC)—Part 4-38: Testing and measurement techniques—Test, verification and calibration protocol for voltage fluctuation and flicker compliance test systems

IEC 61000-4-39 Electromagnetic compatibility (EMC)—Part 4-39: Testing and measurement techniques—Radiated fields in close proximity—Immunity test

Appendix 5

Abbreviations

ABS—American Bureau of Shipping
AC—Alternating Current
ACS—Access Control System
AE—Auxiliary Engine
AFS—International Convention on the Control of Harmful Anti-Fouling Systems
AHU—Air Handling Unit
AI—Artificial Intelligence
AIS—Automatic Identification System
AISC—American Institute of Steel Construction
AMS—Alarm and Monitoring System
AoC—Acknowledge of Compliance
API—American Petroleum Institute
ATEX—Atmospheres Explosives
BNWAS—Bridge Navigational Watch Alarm System
BV—Bureau Veritas
BWM—Ballast Water Management
CAN—Controller Area Network
CEC—Canadian Electrical Code
C&E—Cause and Effect
CCR—Central Control Room
CCS—China Classification Society
CCTV—Close Circuit Television
CMMS—Computerized Maintenance Management System
CPU—Central Processing Unit
CMS—Control and Monitoring System
CMS—Continuous Machinery Survey
CRS—Croatian Register of Shipping
DC—Direct Current
DnVGL—Det Norske Veritas Germanischer Lloyd
DP—Dynamic Positioning

DPU—Distributed Processing Unit
DSS—Decision Support System
EAS—Extension Alarm System
ECC—Engine Control Room
ECR—Engine Control Console
ECRC—Engine Control Room Console
EEMUA—Engineering Equipment and Material Users' Association
EEN—Electrical and Instrumentation Engineer
EG—Emergency Generator
EMC—Electromagnetic Compatibility
ESB—Emergency Switchboard
ESD—Emergency Shutdown
ETO—Electronic Technical Officer
EU—European Union
EUR-LEX—European Union Legislation Webpage
F&G—Fire and Gas
FA—Fire Alarm
FD—Functional Description
FMEA—Failure Modes and Effect Analysis
FPSO—Floating, Production, Storage and Offloading Units
FTC—Field Termination Cabinet
GA—General Alarm
GA—General Arrangement
GLONASS—Global Navigation Satellite System (Russian)
GMDSS—Global Maritime Distress and Safety System
GNSS—Global Navigation Satellite System
GPS—Global Positioning System
FAT—Factory Acceptance Test
FTC—Field Termination Cabinet
FS—Field Station
FSS—International Code for Fire Safety Systems
GA—General Arrangement
HiPAP—High Precision Acoustic Positioning
HC—Hydrocarbon
HMI—Human Machine Interface
HPU—Hydraulic Power Unit
HPR—Hydro acoustic Positioning Reference
HSE—Health and Safety Executive
HSG—Health and Safety Guidance
HVAC—Heating, Ventilation, Air Conditioning
H2S—Hydrogen Sulphide
I/O—Inlet/Outlet
IACS—International Association Classification Societies
IAS—Integrated Automation Systems

ICMS—Integrated Control and Monitoring System
IP—Ingress Protection
ICSS—Integrated Control and Safety System
IEC—Electro-technical Commission
IEEE—Electrical and Electronics Engineers
IGBT—Insulated Gate Bipolar Transistor
IMCA—International Marine Contractors Association
IMO—International Maritime Organization
INMARSAT—International Mobile Satellite Organization
IPS—Information and Planning System
IRS—Indian Register of Shipping
IS—Intact Stability
ISA—International Society of Automation
ISPS—International Ship and Port Facility Security
IS—Intrinsically Safety
ISO—International Organization for Standardization
ITT—Information to Tender
ITT—Information to Tender
LAN—Local Area Network
LCD—Liquid Crystal Display
LCP—Local Control Panel
LECR—Local Engine Control Room
LNG—Liquefied Natural Gas
LO—Lubrication Oil
LRS, LR—Lloyds Register
LSA—Live Saving Appliances
MARPOL—International Convention for the Prevention of Pollution from Ships
MAH—Major Accident Hazard
MASS—Marine Autonomous Surface Ship
MC—Mechanical Completion
MCC—Motor Control Centre
MCP—Manual Call Point
MCT—Multi Cable Transit
MDO—Marine Diesel Oil
ME—Main Engine
MGPS—Marine Growth Protection system
MODU—Mobile Offshore Drilling Unit
MRS—Russian Maritime Register of Shipping
MSB—Main Switchboard
MSC—Marine Safety Committee
NDU—Network Distribution Unit
NEC—National Electrical Code
NK—Nippon Kaiji Kuokai
NMA—Norwegian Maritime Authority

NMA—Norwegian Maritime Authority
NMA—Norwegian Maritime Authority
NORSOK—Norwegian Standardization Organization for Petroleum Industry
NMD—Norwegian Maritime Directorate
NOx—Nitrogen Oxide air pollution
NPD—Norwegian Petroleum Directorate
MEPC—Marine Environment Protection Committee
MLC—Maritime Labour Convention, Marine Environment Protection Committee
MMI—Man Machine Interface
OS (O/S)—Operator Station
P&I—Protection and Indemnity Insurance Clubs
P&ID—Piping and Instrumentation Diagrams
PA—Public Address
PA/GA—Public Address and General Alarm
PC—Personal Computer
PLC—Programmable Logic Controller
PFEER—Prevention of Fire and Explosion, and Emergency Response
P.E.—Protective Earth
PMS—Power Management System
PMS—Planned Maintenance System
PRS—Polish Register of Shipping
PSA—Petroleum Safety Authority
PSPC—Performance Standards for Protective Coating
RAP—Remote Access Platform
RDS—Remote Diagnostic System
RFQ—Request for Quotation
RFQ—Request for Quotation
RINA—Registro Italiano Navale
RINA—Remotely Operated Vehicle
RIO—Remote Inlet/Outlet
RO—Recognized Organization
RPM—Revolution per minute
SAS—Safety and Automation System
SAU—Signal Acquisition Unit
SCADA—Supervisory Control And Data Acquisition
SFI- Classification and numbering system used by maritime and offshore industry
SIL—Safety Integrity Level
SLD—Single Line Diagram
SOLAS—Safety of Life at Sea Convention
SOx—Sulfur Oxide air pollution
TA—Technical Agreement
UKCS—United Kingdom Continental Shelf

UL—Underwriters Laboratory
UMS—Unattended Machinery Space
UN—United Nations
UPS—Uninterrupted Power Supply
UR—Unified Requirement
USCG—United States Coast Guard
VDU—Visual Display Unit
VFD—Variable Frequency Drive
VHF—Very High Frequency
VPN—Virtual Private Network
VRC—Valve Remote Control
VIAS—Wartsila Integrated Automation System
WCM—Watch Calling System
WCU—Watch Call Unit
WI—Work Instruction
WO—Work Order

References

1. Safety of Life at Sea (SOLAS) Convention, International Maritime Organisation (IMO).
2. Code for the Construction and Equipment of Mobile Offshore Drilling Units (MODU Code), International Maritime Organisation (IMO) by International Maritime Organisation (IMO).
3. Rules for Building and Classing Steel Vessels, American Bureau of Shipping (ABS).
4. Rules for Building and Classing Mobile Offshore Drilling Units, American Bureau of Shipping (ABS).
5. Rules for Classification Ships, Det Norske Veritas Germanischer Lloyd (DNVGL).
6. Offshore Standard DNVGL-OS-D202 Automation, Safety and Telecommunication systems, Det Norske Veritas Germanischer Lloyd (DnVGL).
7. Standards IEC 60092 series Electrical Installations in Ships, International Electro-technical Commission.
8. Standards IEC 61892 series Mobile and Fixed Offshore Units—Electrical Installations, International Electro-technical Commission.
9. Standard IEC 60533, Electrical and electronic Installations in Ships—Electromagnetic Compatibility, International Electro-technical Commission.
10. Standards IEC 60079 series, Explosive Atmospheres, International Electro-technical Commission.
11. Standards IEC 61000-4 series Electromagnetic Compatibility (EMC), International Electro-technical Commission.
12. ISO 14617 Graphical Symbols for Diagrams.
13. IEC 62682 Management of Alarm Systems for the Process Industries.
14. EEMUA 191—Alarm Systems—A Guide to Design, Management and Procurement.
15. ISA-18.2 Management of Alarm Systems for the Process Industries.
16. YA-711 Principles for Alarm System Design, Petroleum Safety Authority.
17. IEC-61508 Functional Safety of Electrical/Electronic/Programmable Electronic Safety-Related Systems.
18. IEC-61511 Functional Safety—Safety Instrumented Systems for the Process Industry Sector.

List of Figures

Index

Note: Page numbers followed by *f* indicate figures and *np* indicate footnotes.

Printed in the United States
By Bookmasters